Practical GIS

Use tools such as QGIS, PostGIS, and GeoServer to build
powerful GIS solutions

Gábor Farkas

BIRMINGHAM - MUMBAI

Practical GIS

First published: June 2017

Production reference: 1080617

Published by Packt Publishing Ltd.
Livery Place
35 Livery Street
Birmingham
B3 2PB, UK.

ISBN 978-1-78712-332-8

www.packtpub.com

Credits

Author
Gábor Farkas

Reviewers
Mark Lewin
David Bianco

Commissioning Editor
Aaron Lazar

Acquisition Editor
Angad Singh

Content Development Editor
Lawrence Veigas

Technical Editor
Abhishek Sharma

Copy Editor
Sonia Mathur

Project Coordinator
Prajakta Naik

Proofreader
Safis Editing

Indexer
Mariammal Chettiyar

Graphics
Abhinash Sahu

Production Coordinator
Shantanu Zagade

About the Author

Gábor Farkas is a PhD student in the University of Pécs's Institute of Geography. He holds a master's degree in geography, although he moved from traditional geography to pure geoinformatics in his early studies. He often studies geoinformatical solutions in his free time, keeps up with the latest trends, and is an open source enthusiast. He loves to work with GRASS GIS, PostGIS, and QGIS, but his all time favorite is Web GIS, which mostly covers his main research interest.

About the Reviewer

Mark Lewin has been developing, teaching, and writing about software for over 16 years. His main interest is GIS and web mapping. Working for ESRI, the world's largest GIS company, he acted as a consultant, trainer, course author, and a frequent speaker at industry events. He has subsequently expanded his knowledge to include a wide variety of open source mapping technologies and a handful of relevant JavaScript frameworks including Node.js, Dojo, and JQuery.

Mark now works for Oracle's MySQL curriculum team, focusing on creating great learning experiences for DBAs and developers, but remains crazy about web mapping.

He is the author of books such as *Leaflet.js Succinctly*, *Go Succinctly*, and *Go Web Development Succinctly* for *Syncfusion*. He is also the co-author of the forthcoming second edition of *Building Web and Mobile ArcGIS Server Applications with JavaScript*, which is to be published by Packt.

I would like to thank the production team at Packt for keeping me on schedule, and also my wonderful children who have seen less of me during the process than they would have done otherwise!

www.PacktPub.com

For support files and downloads related to your book, please visit www.PacktPub.com.

Did you know that Packt offers eBook versions of every book published, with PDF and ePub files available? You can upgrade to the eBook version at www.PacktPub.com and as a print book customer, you are entitled to a discount on the eBook copy. Get in touch with us at service@packtpub.com for more details.

At www.PacktPub.com, you can also read a collection of free technical articles, sign up for a range of free newsletters and receive exclusive discounts and offers on Packt books and eBooks.

https://www.packtpub.com/mapt

Get the most in-demand software skills with Mapt. Mapt gives you full access to all Packt books and video courses, as well as industry-leading tools to help you plan your personal development and advance your career.

Why subscribe?

- Fully searchable across every book published by Packt
- Copy and paste, print, and bookmark content
- On demand and accessible via a web browser

Customer Feedback

Thanks for purchasing this Packt book. At Packt, quality is at the heart of our editorial process. To help us improve, please leave us an honest review on this book's Amazon page at `https://www.amazon.com/dp/1787123324`.

If you'd like to join our team of regular reviewers, you can e-mail us at `customerreviews@packtpub.com`. We award our regular reviewers with free eBooks and videos in exchange for their valuable feedback. Help us be relentless in improving our products!

I'm dedicating this book to every open source contributor, researcher, and teacher using and promoting open source technologies. It is your mentality, curiosity, and willingness to put aside proprietary solutions, which makes knowledge and technology more accessible. It is your work that really makes a difference, by letting everyone eager to learn and willing to look under the hood develop themselves.

Table of Contents

Preface

In the past, professional spatial analysis in the business sector was equivalent to buying an ArcGIS license, storing the data in some kind of Esri database, and publishing results with the ArcGIS Server. These trends seem to be changing in the favor of open source software. As FOSS (free and open source software) products are gaining more and more power due to the hard work of the enthusiastic open source GIS community, they pique the curiosity of the business sector at a growing rate. With the increasing number of FOSS GIS experts and consulting companies, both training and documentation--the two determining factors that open source GIS products traditionally lacked--are becoming more available.

What this book covers

Chapter 1, *Setting Up Your Environment*, guides you through the basic steps of creating an open source software infrastructure you can carry out your analyses with. It also introduces you to popular open data sources you can freely use in your workflow.

Chapter 2, *Accessing GIS Data with QGIS*, teaches you about the basic data models used in GIS. It discusses the peculiarities of these data models in detail, and also makes you familiar with the GUI of QGIS by browsing through some data.

Chapter 3, *Using Vector Data Effectively*, shows you how you can interact with vector data in the GIS software. It discusses GUI-based queries, SQL-based queries, and basic attribute data management. You will get accommodated to the vector data model and can use the attributes associated to the vector features in various ways.

Chapter 4, *Creating Digital Maps*, discusses the basics of digital map making by going through an exhaustive yet simple example in QGIS. It introduces you to the concept of projections and spatial reference systems, and the various steps of creating a digital map.

Chapter 5, *Exporting Your Data*, guides you through the most widely used vector and raster data formats in GIS. It discusses the strengths and weaknesses of the various formats, and also gives you some insight on under what circumstances you should choose a particular spatial data format.

Chapter 6, *Feeding a PostGIS Database*, guides you through the process of making a spatial database with PostGIS. It discusses how to create a new database, and how to fill it with various kinds of spatial data using QGIS. You will also learn how to manage existing PostGIS tables from QGIS.

Chapter 7, *A PostGIS Overview,* shows what other options you have with your PostGIS database. It leaves QGIS and talks about important PostgreSQL and PostGIS concepts by managing the database created in the previous chapter through PostgreSQL's administration software, pgAdmin.

Chapter 8, *Spatial Analysis in QGIS,* goes back to QGIS in order to discuss vector data analysis and spatial modeling. It shows you how different geometry types can be used to get some meaningful results based on the features' spatial relationship. It goes through the practical textbook example of delimiting houses based on some customer preferences.

Chapter 9, *Spatial Analysis on Steroids - Using PostGIS,* reiterates the example of the previous chapter, but entirely in PostGIS. It shows how a good software choice for the given task can enhance productivity by minimizing manual labor and automating the entire workflow. It also introduces you to the world of PostGIS spatial functions by going through the analysis again.

Chapter 10, *A Typical GIS Problem,* shows raster analysis, where spatial databases do not excel. It discusses typical raster operations by going through a decision making process. It sheds light on typical considerations related to the raster data model during an analysis, while also introducing some powerful tools and valuable methodology required to make a good decision based on spatial factors and constraints.

Chapter 11, *Showcasing Your Data,* goes on to the Web stack, and discusses the basics of the Web, the client-server architecture, and spatial servers. It goes into details on how to use the QGIS Server to create quick visualizations, and how to use GeoServer to build a powerful spatial server with great capabilities.

Chapter 12, *Styling Your Data in GeoServer,* discusses the basic vector and raster symbology usable in GeoServer. It goes through the styling process by using traditional SLD documents. When the concepts are clear, it introduces the powerful and convenient GeoServer CSS, which is also based on SLD.

Chapter 13, *Creating a Web Map,* jumps to the client side of the Web and shows you how to create simple web maps using the server architecture created before, and the lightweight web mapping library--Leaflet. It guides you through the process of creating a basic web map, ranging from creating an HTML document to scripting it with JavaScript.

Appendix shows additional information and interesting use cases of the learned material through images and short descriptions.

What you need for this book

For this book, you will need to have a computer with mid-class computing capabilities. As the open source GIS software is not that demanding, you don't have to worry about your hardware specification when running the software, although some of the raster processing tools will run pretty long (about 5-10 minutes) on slower machines.

What you need to take care of is that you have administrator privileges on the machine you are using, or the software is set up correctly by an administrator. If you don't have administrator privileges, you need to write the privilege at least to the folder used by the web server to serve content.

Who this book is for

The aim of this book is to carry on this trend and demonstrate how even advanced spatial analysis is convenient with an open source product, and how this software is a capable competitor of proprietary solutions. The examples from which you will learn how to harness the power of the capable GIS software, QGIS; the powerful spatial ORDBMS (object-relational database management system), PostGIS; and the user-friendly geospatial server, GeoServer are aimed at IT professionals looking for cheap alternatives to costly proprietary GIS solutions with or without basic GIS training.

On the other hand, anyone can learn the basics of these great open source products from this practical guide. If you are a decision maker looking for easily producible results, a CTO looking for the right software, or a student craving for an easy-to-follow guide, it doesn't matter. This book presents you the bare minimum of the GIS knowledge required for effective work with spatial data, and thorough but easy-to-follow examples for utilizing open source software for this work.

Conventions

In this book, you will find a number of text styles that distinguish between different kinds of information. Here are some examples of these styles and an explanation of their meaning.

Code words in text, database table names, folder names, filenames, file extensions, pathnames, dummy URLs, and user input are shown as follows: "It uses the * wildcard for selecting everything from the table named `table`, where the content of the column named `column` matches `value`."

A block of code is set as follows:

```
SELECT ST_Buffer(geom, 200) AS geom
 FROM spatial.roads r
 WHERE r.fclass LIKE 'motorway%' OR r.fclass LIKE 'primary%';
```

Any command-line input or output is written as follows:

```
update-alternatives --config java
```

New terms and **important words** are shown in bold. Words that you see on the screen, for example, in menus or dialog boxes, appear in the text like this: "If we open the **Properties** window of a vector layer and navigate to the **Style** tab, we can see the **Single symbol** method applied to the layer."

Warnings or important notes appear in a box like this.

Tips and tricks appear like this.

Reader feedback

Feedback from our readers is always welcome. Let us know what you think about this book—what you liked or disliked. Reader feedback is important for us as it helps us develop titles that you will really get the most out of.

To send us general feedback, simply e-mail feedback@packtpub.com, and mention the book's title in the subject of your message.

If there is a topic that you have expertise in and you are interested in either writing or contributing to a book, see our author guide at www.packtpub.com/authors.

Customer support

Now that you are the proud owner of a Packt book, we have a number of things to help you to get the most from your purchase.

Downloading the example code

You can download the example code files for this book from your account at `http://www.packtpub.com`. If you purchased this book elsewhere, you can visit `http://www.packtpub.com/support`and register to have the files e-mailed directly to you.

You can download the code files by following these steps:

1. Log in or register to our website using your e-mail address and password.
2. Hover the mouse pointer on the **SUPPORT** tab at the top.
3. Click on **Code Downloads & Errata**.
4. Enter the name of the book in the **Search** box.
5. Select the book for which you're looking to download the code files.
6. Choose from the drop-down menu where you purchased this book from.
7. Click on **Code Download**.

Once the file is downloaded, please make sure that you unzip or extract the folder using the latest version of:

- WinRAR / 7-Zip for Windows
- Zipeg / iZip / UnRarX for Mac
- 7-Zip / PeaZip for Linux

The code bundle for the book is also hosted on GitHub at `https://github.com/PacktPublishing/Practical-GIS`. We also have other code bundles from our rich catalog of books and videos available at `https://github.com/PacktPublishing/`. Check them out!

Downloading the color images of this book

We also provide you with a PDF file that has color images of the screenshots/diagrams used in this book. The color images will help you better understand the changes in the output. You can download this file from `https://www.packtpub.com/sites/default/files/downloads/PracticalGIS_ColorImages.pdf`.

Errata

Although we have taken every care to ensure the accuracy of our content, mistakes do happen. If you find a mistake in one of our books—maybe a mistake in the text or the code—we would be grateful if you could report this to us. By doing so, you can save other readers from frustration and help us improve subsequent versions of this book. If you find any errata, please report them by visiting `http://www.packtpub.com/submit-errata`, selecting your book, clicking on the **Errata Submission Form** link, and entering the details of your errata. Once your errata are verified, your submission will be accepted and the errata will be uploaded to our website or added to any list of existing errata under the Errata section of that title.

To view the previously submitted errata, go to `https://www.packtpub.com/books/content/support` and enter the name of the book in the search field. The required information will appear under the **Errata** section.

Piracy

Piracy of copyrighted material on the Internet is an ongoing problem across all media. At Packt, we take the protection of our copyright and licenses very seriously. If you come across any illegal copies of our works in any form on the Internet, please provide us with the location address or website name immediately so that we can pursue a remedy.

Please contact us at `copyright@packtpub.com` with a link to the suspected pirated material.

We appreciate your help in protecting our authors and our ability to bring you valuable content.

Questions

If you have a problem with any aspect of this book, you can contact us at `questions@packtpub.com`, and we will do our best to address the problem.

1
Setting Up Your Environment

The development of open source GIS technologies has reached a state where they can seamlessly replace proprietary software in the recent years. They are convenient, capable tools for analyzing geospatial data. They offer solutions from basic analysis to more advanced, even scientific, workflows. Moreover, there are tons of open geographical data out there, and some of them can even be used for commercial purposes. In this chapter, we will acquaint ourselves with the open source software used in this book, install and configure them with an emphasis on typical pitfalls, and learn about some of the most popular sources of open data out there.

In this chapter, we will cover the following topics:

- Installing the required software
- Configuring the software
- Free geographical data sources
- Software and data licenses

Understanding GIS

Before jumping into the installation process, let's discuss **geographic information systems (GIS)** a little bit. GIS is a system for collecting, manipulating, managing, visualizing, analyzing, and publishing spatial data. Although these functionalities can be bundled in a single software, by definition, GIS is not a software, it is rather a set of functionalities. It can help you to make better decisions, and to get more in-depth results from data based on their spatial relationships.

The most important part of the former definition is spatial data. GIS handles data based on their locations in a coordinate reference system.

This means, despite GIS mainly being used for handling and processing geographical data (data that can be mapped to the surface of Earth), it can be used for anything with dimensions. For example, a fictional land like Middle-Earth, the Milky Way, the surface of Mars, the human body, or a single atom. The possibilities are endless; however, for most of them, there are specialized tools that are more feasible to use.

The functionalities of a GIS outline the required capabilities of a GIS expert. Experts need to be able to collect data either by surveying, accessing an other's measurements, or digitizing paper maps, just to mention a few methods. Collecting data is only the first step. Experts need to know how to manage this data. This functionality assumes knowledge not only in spatial data formats but also in database management. Some of the data just cannot fit into a single file. There can be various reasons behind this; for example, the data size or the need for more sophisticated reading and writing operations. Experts also need to visualize, manipulate, and analyze this data. This is the part where GIS clients come in, as they have the capabilities to render, edit, and process datasets. Finally, experts need to be able to create visualizations from the results in order to show them, verify decisions, or just help people interpreting spatial patterns. This phase was traditionally done via paper maps and digital maps, but nowadays, web mapping is also a very popular means of publishing data.

From these capabilities, we will learn how to access data from freely available data sources, store and manage them in a database, visualize and analyze them with a GIS client, and publish them on the Web.

Setting up the tools

Most of the software used in this book is platform-dependent; therefore, they have different ways of getting installed on different operating systems. I assume you have enough experience with your current OS to install software, and thus, we will focus on the possible product-related pitfalls in a given OS. We will cover the three most popular operating systems--Linux, Windows, and macOS. If you don't need the database or the web stack, you can skip the installation of the related software and jump through the examples using them.

 Make sure you read the OS-related instructions before installing the software if you do not have enough experience with them.

The list of the software stack used in this book can be found in the following thematically grouped table:

Product	Description	Package name (Linux)
Desktop client		
QGIS GRASS	A universal GUI-based GIS client A GIS client, mainly for advanced geospatial analysis and research purposes	qgis grass
Database		
PostgreSQL PostGIS pgAdmin	A universal ORDBMS to store and query data A spatial extension for PostgreSQL A GUI-based client for administrating PostgreSQL databases	postgresql postgis pgadmin3
Web		
Apache QGIS Server GeoServer Leaflet	A general -purpose web server A server application for publishing QGIS projects on the web A feature-packed spatial server with a user-friendly GUI A lightweight web mapping client	apache2 httpd qgis-server

Some of these packages are changeable; you can try them out if you have enough experience or have some time for tinkering. For example, you can use nginx instead of Apache, or you can use the **WAR** (**Web Archive**) version of GeoServer with your Java servlet instead of the platform independent binary. You can also use pgAdmin 4 and any subversion of GRASS 6 or GRASS 7 (or even both of them).

Installing on Linux

Installing the packages on Linux distributions is pretty straightforward. The dependencies are installed with the packages, when there are any. We only have to watch out for three things prior to installing the packages. First of all, the package name of the Apache web server can vary between different distributions. On distros using RPM packages (for example--Fedora, CentOS, and openSUSE), it is called httpd, while on the ones using DEB packages (for example--Debian and Ubuntu), it is called apache2. On Arch Linux, it is simply called apache.

While Arch Linux is far from the best distribution for using GIS, you can get most of the packages from AUR (Arch User Repository).

The second consideration is related to distributions which do not update their packages frequently, like Debian. GeoServer has a hard dependency of a specific JRE (Java Runtime Environment). We must make sure we have it installed and configured as the default. We will walk through the Debian JRE installation process as it is the most popular Linux distribution with late official package updates. Debian Jessie, the latest stable release of the OS when writing these lines, is packed with OpenJDK 7, while GeoServer 2.11 requires JRE 8:

You can check the JRE version of the latest GeoServer version uses at `http://docs.geoserver.org/latest/en/user/installation/index.html`. You can check the JRE version installed on your OS with the terminal command `update-alternatives --list java`.

1. To install OpenJDK 8, we have to enable the Backports repository according to the official Debian guide at `https://wiki.debian.org/Backports`.
2. If the repository is added, we can reload the packages and install the package `openjdk-8-jre`.
3. The next step is to make this JRE the default one. We can do this by opening a terminal and typing the following command:

    update-alternatives --config java

4. The next step is self-explanatory; we have to choose the new default environment by typing its ID and pressing enter.

Make sure to disable the Backports repository by commenting it out in `/etc/apt/sources.list`, or by checking out its checkbox in Synaptic after installing the required packages. It can boycott further updates in some cases.

The last consideration before installing the packages is related to the actual version of QGIS. Most of the distributions offer the latest version in a decent time after release; however, some of them like Debian do not. For those distros, we can use QGIS's repository following the official guide at `http://www.qgis.org/en/site/forusers/alldownloads.html`.

After all things are set, we can proceed and install the required packages. The order should not matter. If done, let's take a look at GeoServer, which doesn't offer Linux packages to install. It offers two methods for Linux: a WAR for already installed Java servlets (such as Apache Tomcat), and a self-containing platform independent binary. We will use the latter as it's easier to set up:

1. Download GeoServer's platform independent binary from `http://geoserver.org/release/stable/`.

 If there is absolutely no way to install OpenJDK 8 on your computer, you can use GeoServer 2.8, which depends on JRE 7. You can download it from `http://geoserver.org/release/2.8.5/`.

2. Extract the downloaded archive. It can be anywhere as long as we have a write permission to the destination.
3. Start GeoServer with its startup script. To do this, we navigate into the extracted archive from a terminal and run `startup.sh` in its `bin` folder with the following command:

```
cd <geoserver's folder>/bin
./startup.sh
```

4. Optionally, we can detach GeoServer from the shell used by the terminal with the startup command `nohup ./startup.sh > /dev/null &`. This way, we can close the terminal. If we would like to shut down GeoServer manually, we can do so by running its `shutdown.sh` script.

 By default, the shell closes every subprocess it started before terminating itself. By using `nohup`, we override this behavior for the GeoServer process, and by using `&` at the end of the command, we fork the process. This way, we regain control over the shell. The `>/dev/null` part prevents `nohup` from logging GeoServer's verbose startup messages.

Installing on Windows

Installing the required software on Windows only requires a few installers as most of the packages are bundled into the OSGeo4W installer.

1. First of all, we have to download the 32-bit installer from `https://trac.osgeo.o rg/osgeo4w/` as this is the only architecture where an OSGeo version of Apache is bundled.

2. Opening the installer, we can choose between different setups. For our cause, we should choose **Advanced Install**. When we reach the **Select Packages** section, we must choose the following packages as a minimum:

   - Desktop--`grass`, `qgis`
   - Web--`apache`, `qgis-server`

3. The next page tells us we don't have to bother with dependencies as the installer selected them for us automatically.

4. The last step can be quite troublesome as there isn't a general solution; we have to configure Apache and QGIS Server if they don't want to collaborate (opening `http://localhost/qgis/qgis_mapserv.fcgi.exe` returns an **Internal Server Error** or it simply cannot be reached). For a good start, take a look at the official tutorial at `http://hub.qgis.org/projects/quantum-gis/wiki/QGIS_Se rver_Tutorial`.

Don't worry if you end up with no solutions, we will concentrate on GeoServer, which runs perfectly on Windows. Just make sure Apache is installed and working (i.e. `http://localhost` returns a blank page or the OSGeo4W default page), as we will need it later.

If you would like to install the 64-bit version of Apache separately, you can find suggestions on compiled 64-bit binaries at `https://httpd.apach e.org/docs/current/platform/windows.html`.

5. The next thing to consider is the PostgreSQL stack. We can download the installer from `https://www.postgresql.org/download/windows/`, where the EnterpriseDB edition comes with a very handy Stack Builder. After the installation of PostgreSQL, we can use it to install PostGIS. We can find PostGIS in the **Spatial Extensions** menu. The default installer comes with pgAdmin 4, while we will use pgAdmin 3 in this book. The two look and feel similar enough; however, if you would like to install the latter, you can download it from `https://www.pgadmin.org/download/pgadmin-3-windows/`.

> The last thing to install is GeoServer, which is such an easy task that we won't discuss it. You can download the installer from `http://geoserver.org/release/stable/`. Make sure you have Java 8 (`https://java.com/en/download/`) before starting it!

Installing on macOS

Installing the software on macOS could be the most complicated of all (because of GRASS). However, thanks to William Kyngesburye, the compiled version of QGIS already contains a copy of GRASS along with other GIS software used by QGIS. In order to install QGIS, we have to download the disk image from `http://www.kyngchaos.com/software/qgis`.

> If you need the GIS software on OS X 10.6 or older, take a look at Kyngesburye's archive at `http://www.kyngchaos.com/software/archive`. Before installing the software, make sure you read his hints and warnings related to the given image.

PostgreSQL and PostGIS are also available from the same site, you will see the link on the left sidebar. pgAdmin, on the other hand, is available from another source: `https://www.pgadmin.org/download/pgadmin-4-macos/`. Finally, the GeoServer macOS image can be downloaded from `http://geoserver.org/release/stable/`, while its dependency of Java 8 can be downloaded from `https://www.java.com/en/download/`.

> If you would like to use pgAdmin 3 instead, or pgAdmin 4 is not supported by your OS, you can download pgAdmin 3 from `https://www.pgadmin.org/download/pgadmin-3-macos/`.

The only thing left is configuring the QGIS Server. As the OS X and macOS operating systems are shipped with an Apache web server, we don't have to install it. However, we have to make some configurations manually due to the lack of the FastCGI Apache module, on which QGIS Server relies. This configuration can be made based on the official guide at h

`ttp://hub.qgis.org/projects/quantum-gis/wiki/QGIS_Server_Tutorial`.

Getting familiar with the software

Congratulations! You're through the hardest part of this chapter. The following step is to make some initial configurations on the installed software to make them ready to use when we need them. First of all, let's open QGIS. At first glance, it has a lot of tools. However, most of them are very simple and self-explanatory. We can group the parts of the GUI as shown in the following image:

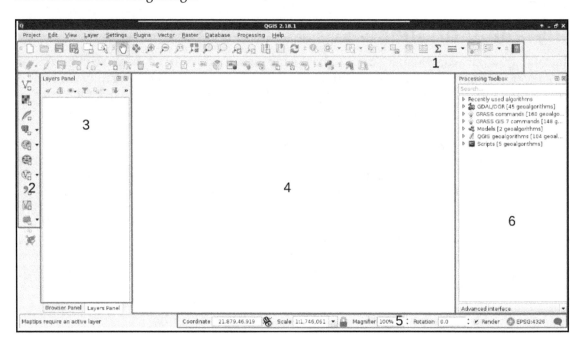

You can learn more about the individual GUI tools from the online QGIS manual at `http://docs.qgis.org/2.14/en/docs/user_manual/introduction/qgis_gui.html`. It is somewhat outdated; however, the GUI hasn't changed much since then (and presumably, it won't change anymore until QGIS 3.0).

We can describe the distinct parts of the QGIS GUI as follows:

1. **Main toolbar**: We can manage our current workflow, pan the map, and make selections and queries from here. Additionally, new tools from plugins will end up somewhere here.
2. **Add layer**: From this handy toolbar, we can add a lot of different spatial data with only a few clicks.
3. **Layer tree**: We can manage our layers from here. We can select them, style them individually, and even apply filters on most of them.
4. **Map canvas**: This is the main panel of QGIS where the visible layers will be drawn. We can pan and zoom our maps with our mouse from here.
5. **Status bar**: These are the simple, yet powerful tools for customizing our view. We can zoom to specified scales, coordinates, and even rotate the map. We can also quickly change our projection, which we will discuss in more depth later.
6. **Processing toolbar**: We can access most of the geoalgorithms bundled in QGIS, and even use other open source GIS clients when they are more fitting for the task.

The only thing we will do now without having any data to display, is customizing the GUI. Let's click on **Settings** and choose the **Options** menu. In the first tab called **General**, we can see some styles to choose from. Don't forget to restart QGIS every time you choose a new style.

 Did you know that like much professional software, QGIS also has a night mode? You can toggle it from the **UI Theme** option in the **General** tab without restarting the software.

The next piece of software we look at is PostGIS via pgAdmin. If we open pgAdmin, the least we will see is an empty **Server Groups** item on the left panel. If this is the case, we have to define a new connection with the plug icon and fill out the form (**Object** | **Create** | **Server** in pgAdmin 4), as follows:

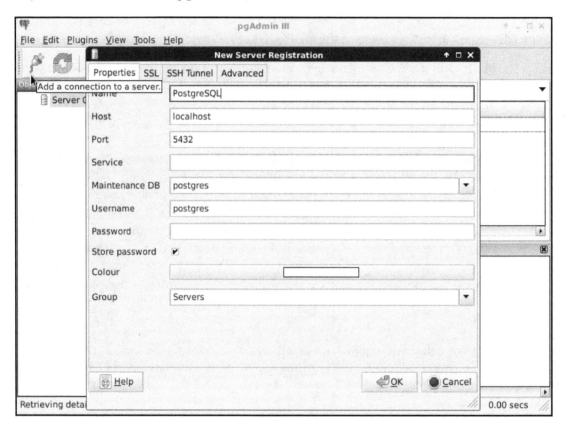

The **Name** can be anything we would like, it only acts as a named item in the list we can choose from. The **Host**, the **Port**, and the **Username,** on the other hand, have to be supplied properly. As we installed PostgreSQL locally, the host is 127.0.0.1, or simply localhost. As the default install comes with the default user postgres (we will refer to users as roles in the future due to the naming conventions of PostgreSQL), we should use that.

On Windows, you can give a password for `postgres` at install time and you can also define the port number. If you changed the default port number or supplied a password, you have to fill out those fields correctly. On other platforms, there is no password by default; however, the database server can only be accessed from the local machine.

Upon connecting to the server, we can see a single database called `postgres`. This is the default database of the freshly installed PostgreSQL. As the next step, we create another database by right-clicking on **Databases** and selecting **New Database**. The database can be named as per our liking (I'm naming it `spatial`). The owner of the database should be the default `postgres` role in our case. The only other parameter we should define is the default character encoding of the database:

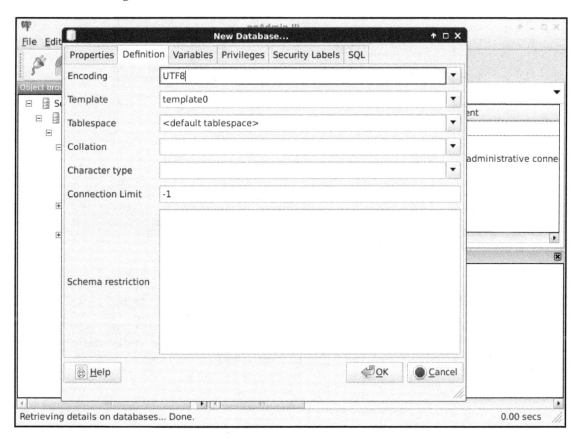

Choosing the **template0** template is required as the default template's character encoding is a simple ASCII. You might be familiar with character encoding; however, refreshing our knowledge a little bit cannot hurt. In ASCII, every character is encoded on 8 bits, therefore, the number of characters which can be encoded is $2^8 = 256$. Furthermore, in ASCII, only the first 7 bits (first 128 places) are reserved, the rest of them can be localized. The first 7 bits (in hexadecimal, 00-7F) can be visualized as in the following table. The italic values show control characters (`https://en.wikipedia.org/wiki/C0_and_C1_control_codes#C`
`0_.28ASCII_and_derivatives.29`):

	0	1	2	3	4	5	6	7
0	*NULL*	*DLE*	␣	0	@	P	`	p
1	*SOH*	*DC1*	!	1	A	Q	a	q
2	*STX*	*DC2*	"	2	B	R	b	r
3	*ETX*	*DC3*	#	3	C	S	c	s
4	*EOT*	*DC4*	$	4	D	T	d	t
5	*ENQ*	*NAK*	%	5	E	U	e	u
6	*ACK*	*SYN*	&	6	F	V	f	v
7	*BELL*	*ETB*	'	7	G	W	g	w
8	*BS*	*CAN*	(	8	H	X	h	x
9	*TAB*	*EM*	)	9	I	Y	I	y
A	*LF*	*SUB*	*	:	J	Z	j	z
B	*VT*	*ESC*	+	;	K	[	k	{
C	*FF*	*FS*	,	<	L	\	l		
D	*CR*	*GS*	-	=	M	]	m	}
E	*SO*	*RS*	.	>	N	^	n	~
F	*SI*	*US*	/	?	O	_	o	*DEL*

 Character encoding is inherited by every table created in the database. As geographic data can have attributes with special local characters, it is strongly recommended to use a UTF-8 character encoding in spatial databases created for storing international data.

About the software licenses

Open source GIS software offer a very high degree of freedom. Their license types can differ; however, they are all permissive licenses. That means we can use, distribute, modify, and distribute the modified versions of the software.

We can also use them in commercial settings and even sell the software if we can find someone willing to buy it (as long as we sell the software with the source code under the same license). The only restriction is for companies who would like to sell their software under a proprietary license using open source components. They simply cannot do that with most of the software, although some of the licenses permit this kind of use, too.

There is one very important thing to watch out for when we use open source software and data. If somebody contributes often years of work to the community, at least proper attribution can be expected. Most of the open source licenses obligate this right of the copyright holder; however, we must distinguish software from data. Most of the licenses of open source software require the adapted product to reproduce the same license agreement. That is, we don't have to attribute the used software in a work, but we must include the original license with the copyright holders' name when we create an application with them. Data, on the other hand, is required to be attributed when we use it in our work.

There are a few licenses which do not obligate us to give proper attribution. These licenses state that the creator of the content waives every copyright and gives the product to the public to use without any restrictions. Two of the most common licenses of this kind are the Unlicense, which is a software license, and the Creative Commons Public Domain, which is in the GIS world mostly used as a data license.

Collecting some data

Now that we have our software installed and configured, we can focus on collecting some open source data. Data collecting (or data capture) is one of the key expertise of a GIS professional and it often covers a major part of a project budget. Surveying is expensive (for example, equipment, amortization, staff, and so on); however, buying data can also be quite costly. On the other hand, there is open and free data out there, which can drastically reduce the cost of basic analysis. It has some drawbacks, though. For example, the licenses are much harder to attune with commercial activity, because some of them are more restrictive.

There are two types of data collection. The first one is primary data collection, where we measure spatial phenomena directly. We can measure the locations of different objects with GPS, the elevation with radar or lidar, the land cover with remote sensing. There are truly a lot of ways of data acquisition with different equipment. The second type is secondary data collection, where we convert already existing data for our use case. A typical secondary data collection method is digitizing objects from paper maps. In this section, we will acquire some open source primary data.

If you do not feel like downloading anything from the following data sources, you can work with the sample dataset of this book. The sample covers Luxembourg, therefore you can download and visualize it in no time.

The only thing to consider is our study area. We should choose a relatively small administrative division, like a single county. For example, I'm choosing the county I live in as I'm quite familiar with it and it's small enough to make further analysis and visualization tasks fast and simple:

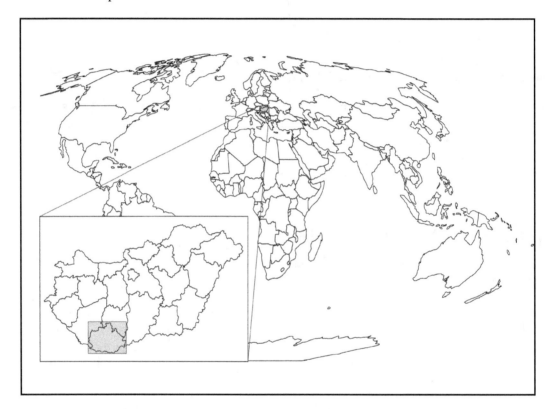

 Make sure you create a folder for the files that we will download. You should extract every dataset in a different folder with a talkative name to keep a clean working directory and to ease future work.

Getting basic data

The first data we will download is the administrative boundaries of our country of choice. Open data for administrative divisions are easy to find for the first two levels, but it becomes more and more scarce for higher levels. The first level is always the countries' boundaries, while higher levels depend on the given country. There is a great source for acquiring the first three levels for every country in a fine resolution: GADM or Global Administrative Areas. We will talk about administration levels in more details in a later chapter. Let's download some data from `http://www.gadm.org/country` by selecting our study area, and the file format as **Shapefile**:

In the zipped archive, we will need the administrative boundaries, which contain our division of choice. If you aren't sure about the correct dataset, just extract everything and we will choose the correct one later.

The second vector dataset we download is the GeoNames archive for the country encasing our study area. GeoNames is a great place for finding data points. Every record in the database is a single point with a pair of coordinates and a lot of attribute data. Its most instinctive use case is for geocoding (linking names to locations). However, it can be a real treasure box for those who can link the rich attribute data to more meaningful objects. The country-level data dumps can be reached at `http://download.geonames.org/export/dump` `/` through the countries' two-letter ISO codes.

ISO (**International Organization of Standards**) is a large-scale organization maintaining a lot of standards for a wide variety of use cases. Country names also have ISO abbreviations, which can be reached at `http://www.geonames.org/countries/` in the form of a list. The first column contains the two-letter ISO codes of the countries.

Licenses

GADM's license is very restrictive. We are free to use the downloaded data for personal and research purposes but we cannot redistribute it or use it in commercial settings. Technically, it isn't open source data as it does not give the four freedoms of using, modifying, redistributing the original version, and redistributing the modified version without restrictions. That's why the example dataset doesn't contain GADM's version of Luxembourg.

There is another data source, called Natural Earth, which is truly open source but it offers data only for the first two levels and on a lower resolution. If you need some boundaries with the least effort, make sure you check it out at `http://www.naturalearthdata.com/downloads/`.

GeoNames has two datasets--a commercially licensed premium dataset and an open source one. The open source data can be used for commercial purposes without restrictions.

Accessing satellite data

Data acquisition with instruments mounted on airborne vehicles is commonly called remote sensing. Mounting sensors on satellites is a common practice by space agencies (for example, NASA and ESA), and other resourceful companies. These are also the main source of open source data as both NASA and ESA grant free access to preprocessed data coming from these sensors. In this part of the book, we will download remote sensing data (often called imagery) from USGS's portal: Earth Explorer. It can be found at `https://earthexplorer.usgs.gov/`. As the first step, we have to register an account in order to download data.

If you would like to download Sentinel-2 data instead of Landsat imagery, you can find ESA's Copernicus data portal at `https://scihub.copernicus.eu/`.

When we have an account, we should proceed to the Earth Explorer application and select our study area. We can select an area on the map by holding down the *Shift* button and drawing a rectangle with the mouse, as shown in the following screenshot:

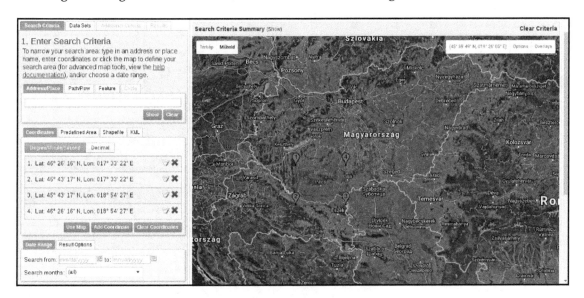

Active remote sensing

As the next step, we should select some data from the **Data Sets** tab. There are two distinct types of remote sensing based on the type of sensor: active and passive. In active remote sensing, we emit some kind of signal from the instrument and measure its reflectance from the target surface. We make our measurement from the attributes of the reflected signal. Three very typical active remote sensing instruments are **radar** (**radio detection and ranging**) using radio waves, **lidar** (**light detection and ranging**) using laser, and **sonar** (**sound navigation and ranging**) using sound waves. The first dataset we download is **SRTM** (**Shuttle Radar Topographic Mission**), which is a **DEM** (**digital elevation model**) produced with a radar mounted on a space shuttle.

For this, we select the **Digital Elevation** item and then **SRTM**. Under the **SRTM** menu, there are some different datasets from which we need the **1 Arc-Second Global**. Finally, we push the **Results** button, which navigates us to the results of our query. In the results window, there are quite a few options for every item, as shown in the following screenshot:

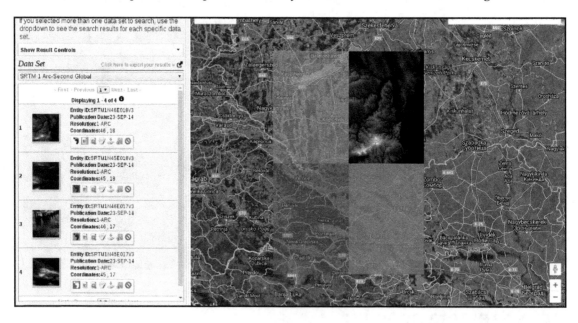

The first two options (**Show Footprint** and **Show Browse Overlay**) are very handy tools to show the selected imagery on the map. The footprint only shows the enveloping rectangle of the data, therefore, it is fast. Additionally, it colors every footprint differently, so we can identify them easily. The overlay tool is handy for getting a glance at the data without downloading it.

Finally, we download the tiles covering our study area. We can download them individually with the item's fifth option called **Download Options**. This offers some options from which we should select the BIL format as it has the best compression rate, thus, our download will be fast.

If you have access to lidar data in your future work, don't hesitate to use it. Up to this time, it offers the most accurate results.

Passive remote sensing

Let's get back to the **Data Sets** tab and select the next type of data we need to download--the Landsat data. These are measured with instruments of the other type--passive remote sensing. In passive remote sensing, we don't emit any signal, just record the electromagnetic radiance of our environment. This method is similar to the one used by our digital cameras except those record only the visible spectrum (about 380-450 nanometers) and compose an RGB picture from the three visible bands instantly. The Landsat satellites use radiometers to acquire multispectral images (bands). That is, they record images from spectral intervals, which can penetrate the atmosphere, and store each of them in different files. There is a great chart created by NASA (`http://landsat.gsfc.nasa.gov/sentinel-2a-launches-our-compliments-our-complements/`) which illustrates the bands of Landsat 7, Landsat 8, and Sentinel-2 along with the atmospheric opacity of the electromagnetic spectrum:

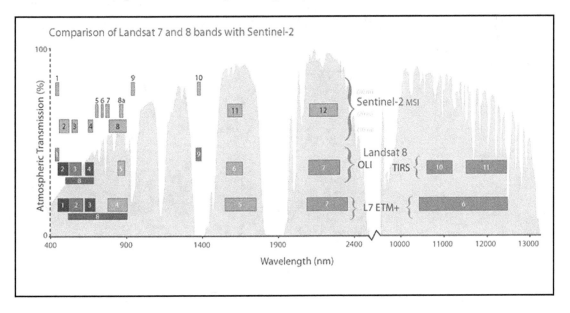

From the **Landsat Archive**, we need the **Pre-Collection** menu. From there, we select **L8 OLI/TIRS** and proceed to the results. With the footprints of the items, let's select an image which covers our study area. As Landsat images have a significant amount of overlap, there should be one image which, at least, mostly encases our study area. There are two additional information listed in every item--the row number and the path number. As these kinds of satellites are constantly orbiting Earth, we should be able to use their data for detecting changes. To assess this kind of use case (their main use case), their orbits are calculated so that, the satellites return to the same spot periodically (in case of Landsat, 18 days). This is why we can classify every image by their path and row information:

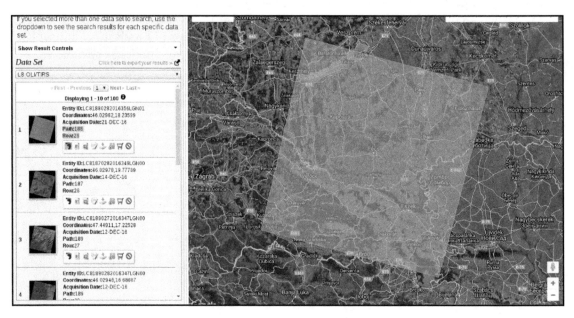

To make sure the images are illuminated the same way every time on a given path/row, this kind of satellite is set on a Sun-synchronous orbit. This means, they see the same spot at the same solar time in every pass. There is a great video created by NASA visualizing Landsat's orbit at `http s://www.youtube.com/watch?v=P-1bujsVa2M`.

Let's note down the path and row information of the selected imagery and go to the **Additional Criteria** tab. We feed the path and row information to the **WRS Path** and **WRS Row** fields and go back to the results. Now the results are filtered down, which is quite convenient as the images are strongly affected by weather and seasonal effects. Let's choose a nice imagery with minimal cloud coverage and download its **Level 1 GeoTIFF Data Product**. From the archive, we will need the TIFF files of bands 1-6.

The `tar.gz` extension is a shorthand for a gzipped tape archive. It is by far the most common compressed archive type on Unix-like operating systems and any decent compressing software can handle it.

Licenses

SRTM is in the public domain; therefore, it can be used without restrictions, and giving attribution is also optional. Landsat data is also open source; however, based on USGS's statement (`https://landsat.usgs.gov/are-there-any-restrictions-use-or-redistri bution-landsat-data`), proper attribution is recommended.

Using OpenStreetMap

The last dataset we put our hands on is the swiss army knife of open source GIS data. OpenStreetMap provides vector data with a great global coverage coming from measurements of individual contributors. OpenStreetMap has a topological structure; therefore, it's great for creating beautiful visualizations and routing services. On the other hand, its collaborative nature makes accuracy assessments hard. There are some studies regarding the accuracy of the whole data, or some of its subsets, but we cannot generalize those results as accuracy can greatly vary even in small areas.

One of the main strengths of OpenStreetMap data is its large collection and variety of data themes. There are administrative borders, natural reserves, military areas, buildings, roads, bus stops, even benches in the database. Although its data isn't surveyed with geodesic precision, its accuracy is good for a lot of cases: from everyday use to small-scale analysis where accuracy in the order of meters is good enough (usually, a handheld GPS has an accuracy of under 5 meters). Its collaborative nature can also be evaluated as a strength as mistakes are corrected rapidly and the content follows real-world changes (especially large ones) with a quick pace.

Accessing OpenStreetMap data can be tricky. There are some APIs and other means to query OSM, although either we need to know how to code or we get everything in one big file. There is one peculiar company which creates thematic data extracts from the actual content--Geofabrik. We can reach Geofabrik's download portal at `http://download.geofab rik.de/`. It allows us to download data in OSM's native **PBF** format (**Protocolbuffer Binary Format**), which is great for filling a PostGIS database with OSM data from the command line on a Linux system but cannot be opened with a desktop GIS client.

It also serves XML data, which is more widely supported, but the most useful extracts for us are the shapefiles.

There are additional providers creating extracts from the OpenStreetMap database. For example, Mapzen's Metro Extracts service can create full extracts for a user-defined city sized area. You just have to register, and use the service at https://mapzen.com/data/metro-extracts/. You might need additional tools, out of the scope of this book, to effectively use the downloaded data though.

Due to various reasons, open source shapefiles are only exported by Geofabrik for small areas. We have to narrow down our search by clicking on links until the shapefile format (**.shp.zip**) is available. This means country-level extracts for smaller countries and regional extracts for larger or denser ones. The term dense refers to the amount of data stored in the OSM database for a given country. Let's download the shapefile for the smallest region enveloping our study area:

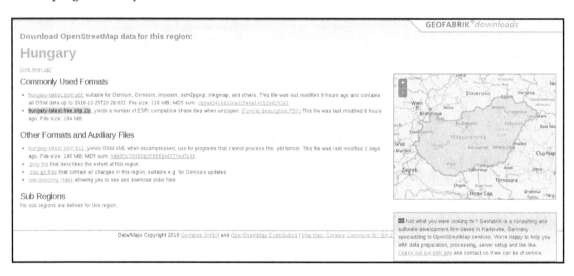

OpenStreetMap license

OpenStreetMap data is licensed under ODbL, an open source license, and therefore gives the four basic freedoms. However, it has two important conditions. The first one is obligatory attribution, while the second one is a share-alike condition. If we use OpenStreetMap data in our work, we must share the OSM part under an ODbL-compatible open source license.

ODbL differentiates three kind of products: collective database, derived database, and produced work. If we create a collective database (a database which has an OSM part), the share-alike policy only applies on the OSM part. If we create a derived database (make modifications to the OSM database), we must make the whole thing open source. If we create a map, a game, or any other work based on the OSM database, we can use any license we would like to. However, if we modify the OSM database during the process, we must make the modifications open source.

 If the license would only have these rules, it could be abused in infinitesimal ways. Therefore, the full license contains a lot more details and some clauses to avoid abuses. You can learn more about ODbL at `htt ps://wiki.osmfoundation.org/wiki/Licence`.

Summary

In this chapter, we installed the required open source GIS software, configured some of them, and downloaded a lot of open source data. We became familiar with open source products, licenses, and data sources. Now we can create an open source GIS working environment from zero and acquire some data to work with. We also gained some knowledge about data collection methods and their nature.

In the next chapter, we will visualize the downloaded data in QGIS. We will learn to use some of the most essential functionalities of a desktop GIS client while browsing our data. We will also learn some of the most basic attributes and specialities of different data types in GIS.

2
Accessing GIS Data With QGIS

Despite the fact that some of the advanced GIS software suggest, we only need to know which buttons to press in order to get instant results, GIS is much more than that. We need to understand the basic concepts and the inner workings of a GIS in order to know the kind of analyses we can perform on our data. We must be able to come up with specific workflows, models which get us the most meaningful results. We also need to understand the reference frame of GIS, how our data behaves in such an environment, and how to interpret those results. In this chapter, we will learn about GIS data models by browsing our data in QGIS, and getting acquainted with its GUI.

In this chapter, we will cover the following topics:

- Graphical User Interface of QGIS
- Opening spatial data in QGIS
- GIS data models

Accessing raster data

The first data type that we will use is raster data. It might be the most familiar to you, as it resembles traditional images. First of all, let's open QGIS. In the browser panel, we can immediately see our downloaded data if we navigate to our working directory. We can easily distinguish vector data from raster data by their icons.

Raster layers have a dedicated icon of a 3x3 pixels image, while vector layers have an icon of a concave polygon:

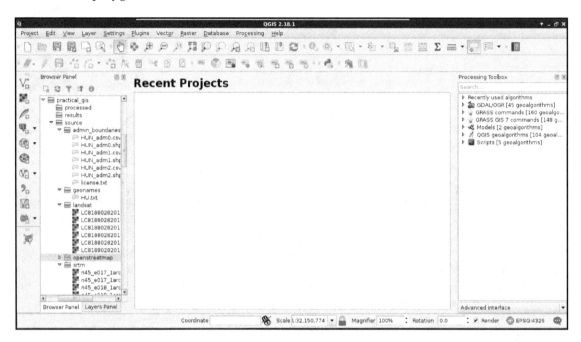

> Don't have a browser panel? You can toggle panels from the **View** menu's **Panels** option. If it is displayed, you can dock it anywhere by dragging it out from its current place and placing it in another part of the GUI.

We can drag and drop most of the data from the browser panel or, alternatively, use the **Add Raster Layer** button from the **Add layer** toolbar and browse the layer. The browser panel is more convenient for easily recognizable layers as it only lists the files we can open and hides auxiliary files with every kind of metadata. Let's drag one of the SRTM rasters to the canvas (or open one with **Add Raster Layer**). This is a traditional, single-band raster.

It is displayed as a greyscale image with the minimum and maximum values displayed in the **Layers Panel**:

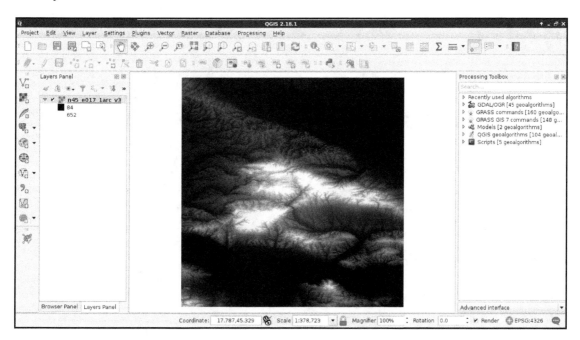

You can save your project any time with the **Save** and **Save As** buttons. The resulting file only contains general data about your current work (for example, paths to your opened layers, styles, and so on) from which you can restore it later. For more information, open a project file in a text or code editor.

As you can see in the preceding screenshot, there is a regular grid with cells painted differently, just like an image. However, based on the maximum value of the data, its colors aren't hard coded into the file, like in an image. Furthermore, it has only a single band, not three or four bands for RGB(A). Let's examine the raster more carefully by zooming in until we can see individual cells.

We can also query them for their values with the **Identify Features** tool by clicking on a cell (raster):

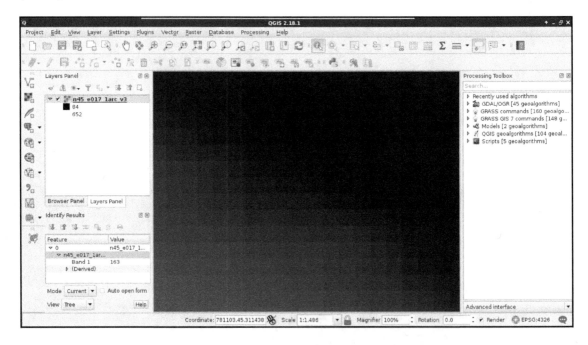

As you can see, we get a number for every cell, which can be quite out of the range of 0-255 representing color codes. These numbers seem arbitrary, and indeed, they are arbitrary. They usually represent some kind of real-world phenomenon, like in our case, the elevation from the mean sea level in meters.

Raster data model

These are the basic properties of the raster data model. Raster data are regular grids (matrices) made up from individual cells with some arbitrary values describing something. The values are only limited by the type of the storage. They can be in the range of bytes, 8-bit integers, 16-bit integers, floating point numbers, and so on. Rasters are always rectangular (like an image); however, they can give a feeling of having some other shape with a special kind of value: NULL or No-Data.

In most of the sophisticated GIS software, there is a special No-Data type for NULL values. However, there is no consensus on how to encode those values. Because of this, it is fairly common to encode NULL values with a number. The chosen No-Data value is often documented and stored in the raster data (if the format permits).

One of the most useful properties of raster data is that their coverage is continuous, while their data can change. They cover their entire extent with coincident cells. If we need a full and continuous coverage (that is, we need a value for every point describing a dynamically changing phenomenon), raster is an obvious choice. On the other hand, they have a fixed layout inherited from their resolution (the size of each cell). There are the following two implications from this property:

- First, the accuracy of a raster is not constant. It covers uniform areas in a given projection. Therefore, on the globe, the area covered by a single raster inherits the distortion of the projection.
- Secondly, if we increase the resolution, the size of the raster data shows a quadratic growth as we have to increase the number of cells in each dimension:

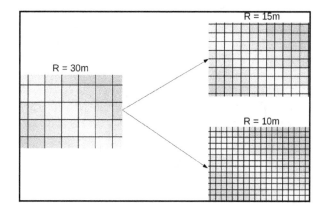

Of course, this property works in two ways. Rasters (especially with square cells) are generally easy and fast to visualize as they can be displayed as regular images. To make the visualization process even faster, QGIS builds pyramids from the opened rasters.

Using pyramids is a computer graphics technique adopted by GIS. Pyramids are downsampled (lower resolution) versions of the original raster layer stored in memory, and are built for various resolutions. By creating these pyramids in advance, QGIS can skip most of the resampling process on lower resolutions (zoom levels), which is the most time-consuming task in drawing rasters.

The last important property of a raster layer is its origin. As raster data behaves as two-dimensional matrices, it can be spatially referenced with only a pair of coordinates. These coordinates, unlike in graphics, are the lower left ones of the data. Let's see what QGIS can tell us about our raster layer. We can see its metadata by right-clicking on it in the **Layers Panel**, choosing **Properties**, and clicking on the **Metadata** tab:

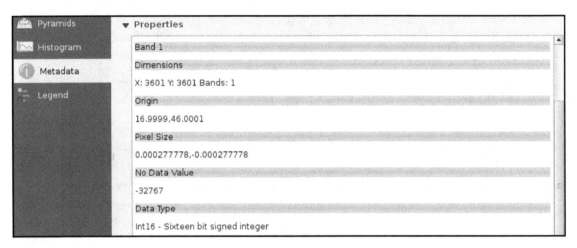

As you can see in the preceding screenshot, our raster layer has a number of rows and columns, one band, an **Origin**, a resolution (**Pixel Size**), a **No-Data value**, and a **Data Type**.

To sum up, the raster data model offers continuous coverage for a given extent with dynamically changing, discrete values in the form of a matrix. We can easily do matrix operations on rasters, but we can also convert them to vectors if it is a better fit for the analysis. The raster data model is mainly used when the type of the data desires it (for example, mapping continuous data, like elevation or terrain, weather, or temperature) and when it is the appropriate model for the measuring instrument (for example, aerial or satellite imaging).

Rasters are boring

To put it simply--absolutely not. Well, maybe in QGIS a little bit, but rasters have potential far beyond the needs of an average GIS analysis. First of all, rasters do not need to be in two-dimensional space. There are 3D rasters called voxels, which can be analyzed in their volume or cut to slices, visualized in a whole, in slices, or as isosurfaces for various values (*Appendix 1.1*).

Furthermore, cells don't have to be squares. It is common practice to have different resolutions in different dimensions. Rasters with rectangular cells are supported by QGIS, and many other open source GIS clients. Rasters don't even need to have four sides. The distortions (we can call it sampling bias in some cases, mostly in statistics) caused by four-sided raster cells can be minimized with hexagons, regular shapes with the most sides capable of a complete coverage:

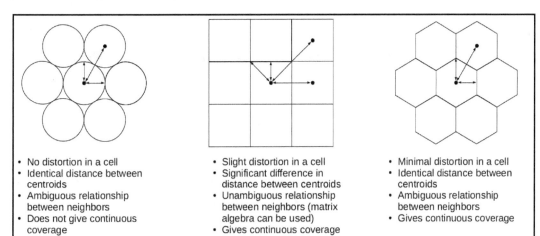

- No distortion in a cell
- Identical distance between centroids
- Ambiguous relationship between neighbors
- Does not give continuous coverage

- Slight distortion in a cell
- Significant difference in distance between centroids
- Unambiguous relationship between neighbors (matrix algebra can be used)
- Gives continuous coverage

- Minimal distortion in a cell
- Identical distance between centroids
- Ambiguous relationship between neighbors
- Gives continuous coverage

 Hexagonal tiling is often implemented by GIS software. However, they usually use a vector data structure to store and handle the hexagonal grid.

Okay, but can rasters only store a single thematic? No, rasters can have multiple bands, and we can even combine them to create RGB visualizations. Finally, as we contradicted almost every **rule** of the raster data model, do individual cells need to coincide (have the same resolution)? Well, technically, yes, but guess what? There are studies about a multi-resolution image format, which can store rasters with different sizes in the same layer. It's now only a matter of professional and business interest to create that format.

Accessing vector data

Now that you've learned the basics of raster data, let's examine vector data. This is the other fundamental data type which is used in GIS. Let's get some vector data at the top of that srtm layer. From the **Browser Panel**, we open up the administrative boundaries layer (the one with the shp extension) containing our study area, and the waterways and traffic layers from the OpenStreetMap data. We can also use the **Add Vector Layer** button from the side toolbar:

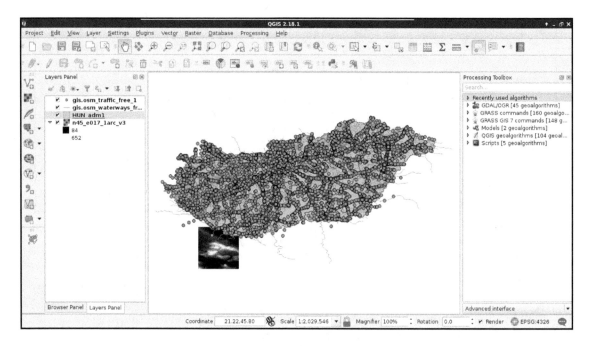

If you don't see the points after opening the traffic data, you might have just opened the traffic areas layer. Geofabrik extracts distinguishable areas from lines and points by appending an a to the file name. The traffic_a_free_1.shp file contains polygons related to traffic (for example, parking lots), while the one named traffic_free_1.shp contains points. You can remove obsolete layers by right-clicking them in the **Layers Panel** and selecting **Remove**.

Now there are three vector layers with three different icons and representation types on our canvas. These are the three main vector types we can work with:

- **Points**: Points are used to represent single punctual occurrences of phenomena
- **Lines**: Lines are used to show linear features and boundaries
- **Polygons**: Polygons are used to delimit areas

If we zoom around, we can see that unlike the raster layer, these layers do not pixelate. They remain as sharp as on lower zoom levels no matter how far we zoom in. Furthermore, if we use the **Identify Features** tool and click on a **feature**, it gets selected. We can see with this that our vector layers consist of arbitrary numbers of points, lines, and irregular shapes:

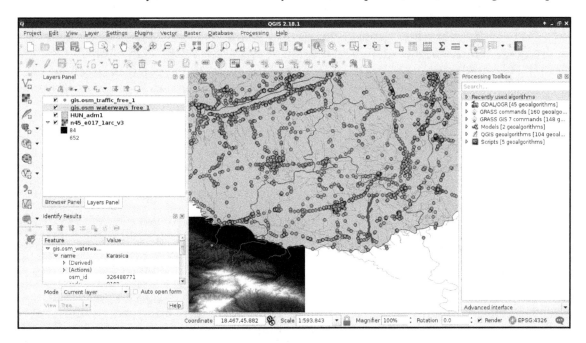

 You can only select features with the **Identify Features** tool from the selected layer. If you wish to query another type of feature, first select its layer from the **Layers Panel**.

Vector data model

Vector data, unlike raster data, does not have a fixed layout. It can be represented as sets of coordinate pairs (or triplets or quads if they have more than two dimensions). The elementary unit of the vector data model is the feature. A feature represents one logically coherent real-world object--an entity. What we consider an entity depends on our needs. For example, if we would like to analyze a forest patch, we gather data from individual trees (entities) and represent them as single points (features). If we would like to analyze land cover, the whole forest patch can be the entity represented by a feature with a polygon geometry. We should take care not to mix different geometry types in a single layer though. It is permitted in some GIS software; however, as some of the geoprocessing algorithms only work on specific types, it can ruin our analysis.

There are more geometry types than points, line strings, and polygons. However, they all rely on these basic types. For example, there are multipart variations of every type. A single multipolygon feature contains an arbitrary number of polygon geometries (e.g. Japan as a single feature), and the like.

Geometry is only part of a vector feature. To a single feature, we can add an arbitrary number of attributes. There are two typical distinct types of attributes--numeric and character string. Different GIS software can handle different subtypes, like integers, floating point numbers, or dates (which can be a main type in some GIS). From these attributes, the GIS software creates a consistent table for every vector layer, which can be used to analyze, query, and visualize features. This attribute table has rows representing features and typed columns representing unique attributes. If at least one feature has a given attribute, it is listed as a whole column with NULL values (or equivalent) in the other rows. This is one of the reasons we should strive for consistency no matter if the used GIS software forces it.

Let's open an attribute table by right-clicking on a vector layer in the **Layers Panel** and selecting **Open Attribute Table**. If we open the **Conditional formatting** panel in the attribute table, we can also see the type of the columns, as shown in the following screenshot:

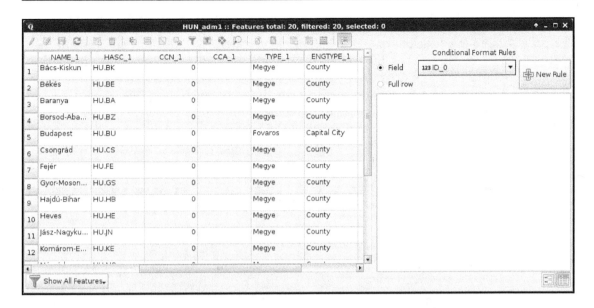

	NAME_1	HASC_1	CCN_1	CCA_1	TYPE_1	ENGTYPE_1
1	Bács-Kiskun	HU.BK	0		Megye	County
2	Békés	HU.BE	0		Megye	County
3	Baranya	HU.BA	0		Megye	County
4	Borsod-Aba...	HU.BZ	0		Megye	County
5	Budapest	HU.BU	0		Fovaros	Capital City
6	Csongrád	HU.CS	0		Megye	County
7	Fejér	HU.FE	0		Megye	County
8	Gyor-Moson...	HU.GS	0		Megye	County
9	Hajdú-Bihar	HU.HB	0		Megye	County
10	Heves	HU.HE	0		Megye	County
11	Jász-Nagyku...	HU.JN	0		Megye	County
12	Komárom-E...	HU.KE	0		Megye	County

 If the same attribute column contains different types for different features, the usual behavior of the GIS software is to set the type of the column to the lowest common denominator. For example, if there are strings and numbers mixed under the same attribute name, the whole column will be string, as strings cannot be treated as numbers, but numbers can be treated as strings.

To sum up, with the vector data model, we can represent entities with shapes consisting of nodes (start and end points) and vertices (mid points) which are coordinates. We can link as many attributes to these geometries as we like (or as the data exchange format permits). The model implies that we can hardly store gradients, as it is optimized to store discrete values associated with a feature. It has a somewhat constant accuracy as we can project the nodes and vertices one by one. Furthermore, the model does not suffer from distortions unlike the raster data model.

As vector geometries do not have a fixed layout, we can edit our features. Let's try it out by selecting the administrative boundaries layer in the **Layers Panel** and starting an edit session with the **Toggle Editing** tool.

 In the following task, we will temporarily ruin our layer. If you are worried about overwriting it with bad geometries, create a backup copy of the affected files first.

We can see every node and vertex of our layer. We can modify these points to reshape our layer. Let's zoom in a little bit to see the individual vertices. From the now enabled editing tools, we select the **Node Tool**. If we click in a polygon with this tool, we can see its vertices highlighted and a list of numbers in the left panel. These are the coordinates of the selected geometry. We can move vertices and segments by dragging them to another part of the canvas. If we move some of the vertices from the neighboring polygon, we can see a gap appearing. This naive geometry model is called the Spaghetti model. Every feature has their sets of vertices individually and there isn't any relationship between them. Consider the following screenshot:

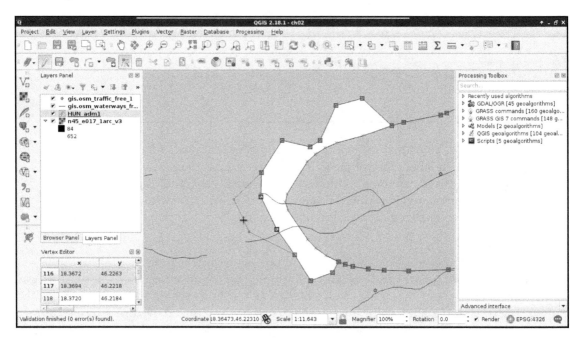

Finally, let's stop the edit session by clicking on the **Toggle Editing** button again. When it asks about saving the modifications, we should choose **Stop without Saving** and QGIS automatically restores the old geometries.

Vector topology - the right way

More complex geometries have more theoretical possibilities, which leads to added complexity. Defining a point is unequivocal, that is, it has only one coordinate tuple. Multi-points and line strings are neither much more complex--they consist of individual and connected coordinate tuples respectively.

Polygons, on the other hand, can contain holes, the holes can contain islands (fills), and theoretically, these structures can be nested infinitesimally. This structure adds a decent complexity for a GIS software. For example, QGIS only supports polygons to the first level-- with holes.

The real complexity, however, only comes with topology. Different features in a layer can have relationships with each other. For example, in our administrative boundary layer, polygons should share borders. They shouldn't have gaps and overlaps. Another great example is a river network. Streams flow into rivers, rivers flow into water bodies. In a vector model, they should be connected like in the real world.

The topological geometry model (or vector model) is the sophisticated way to solve these kinds of relational problems. In this model, points are stored as nodes while other geometries form a hierarchical structure. Line segments contain references to nodes, line strings, and polygons consist of references to segments. By using this hierarchical structure, we can easily handle relationships. This way, if we change the position of a node, every geometry referring to the node changes. Take a look at the following screenshot:

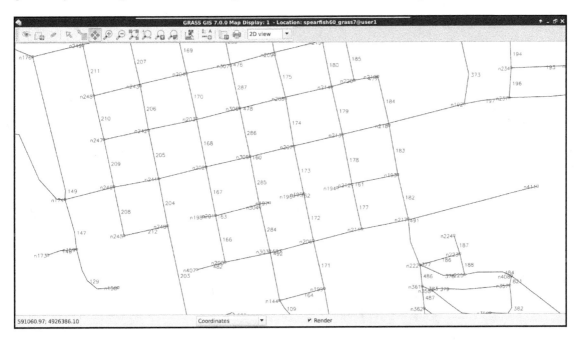

Not every GIS software enforces a topological model. For example, in QGIS, we can toggle topological editing. Let's try it out by checking **Enable topological editing** in **Settings | Snapping Options**. If we edit the boundaries layer again, we can see the neighboring feature's geometry following our changes.

While QGIS does not enforce a topological model, it offers various tools for checking topological consistency. One of the tools is the built-in **Topology Checker** plugin. We can find the tool under the **Add layer** buttons.

If you don't see the tool, it might be disabled in your version of QGIS by default. You can enable it from **Plugins | Manage and Install Plugins**. Go to the **Installed** tab and check the **Topology Checker**.

If we activate the tool, a new panel appears docked under the **Processing Toolbox**. By clicking on **Configure**, we can add some topology rules to the opened vector layers. Let's add two simple rules to the administrative boundaries layer--they must not have gaps or overlaps. Consider the following screenshot:

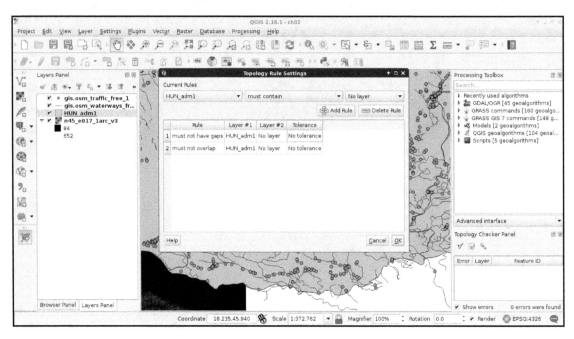

The only thing left to do is to click on the **Validate All** or **Validate Extent** button. If we have some errors, we can navigate between them by clicking on the items one by one.

> If you have some errors, be sure to review them as the tool can create false positives. If you would like to repair some of the real topological errors automatically, you can try out the **Geometry Checker** built-in plugin. You have to enable it first, then you can find the tool at **Vector | Geometry Tools | Check Geometries**. Note that it won't resolve false positives as they are not real errors.

Opening tabular layers

The vector layers we opened so far were dedicated vector data exchange formats; therefore, they had every information coded in them needed for QGIS to open them. There are some cases when we get some data in a tabular format, like in a spreadsheet. These data usually contain points as coordinates in columns and attributes in other columns. They do not store any metadata about the vectors, which we have to gather from readme files, or the team members producing the data.

QGIS can handle one tabular format--**CSV (Comma Separated Values)**, which is an ASCII file format, a simple text file containing tabular data. Every row is in a new line, while fields are separated with an arbitrary field separator character (the default one is the comma). The layout of such a layer is custom; therefore, we need to supply the required information about the table to QGIS. If we try to drag our GeoNames layer to the canvas, QGIS yields to a **Layer is not valid** error.

To open these files, we need to use the **Add Delimited Text Layer** tool from the sidebar (comma icon):

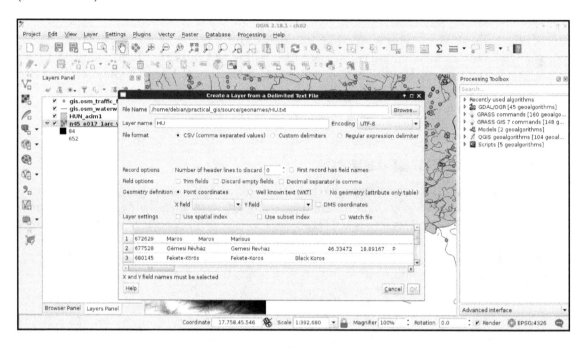

If we browse our **GeoNames** table, we can see that QGIS automatically creates a preview from the accessed data. We can also see that the preview is far from correct. It's time to consult the readme file coming with the GeoNames extract. In the first few lines, we can see the most important information--**The data format is tab-delimited text in utf8 encoding**. Let's select **Custom delimiters** and check **Tab** as a delimiter. Now we only need to supply the columns containing the coordinates. We can see there are no headers in the data. However, as the column descriptions are ordered in the readme file, we can conclude that the fifth column contains the latitude data (**Y field**), while the sixth column contains the longitude data (**X field**). This is the minimum information we can add the layer with:

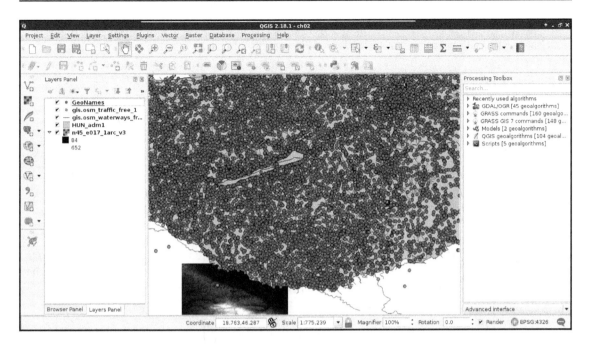

If you have many layers, giving them talkative names can aid you in your work. You can rename a layer by right-clicking on it in the **Layers Panel** and selecting **Rename**. If you have difficulties with rendering the number of points in the GeoNames extract, you can check out the **Render** option on the status bar, load the layer, check out its visibility in the **Layers Panel**, then check in **Render** again. This might make no sense at this point but it will come in handy later.

Understanding map scales

When zooming around the map, we could notice the **Scale** changing in the status bar. GIS software (apart from web mapping solutions) usually use scales instead of zoom levels. The map scale is an important concept of cartography, and its use was inherited by GIS software. The scale shows the ratio (or representative fraction) between the map and the real world. It is a mapping between two physical units:

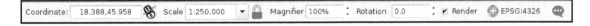

For example, a **Scale** of **1:250,000** means 1 centimeter on the map is 2500 meters (250,000 centimeters) in the real world. However, as the map scale is unitless, it also means 1 inch on the map is 250,000 inches in the real world, and so on. With the scale of the map, we can make explicit statements about its coverage and implicit statements about its accuracy. Large scale maps (for example, with a scale of 1:10,000) cover smaller areas with greater accuracy than medium scale maps (for example, with a scale of 1:500,000), which cover smaller areas with greater accuracy than small scale maps (for example, with a scale of 1:1,000,000).

 Note that the classification of the scales are the inverse of the ratio numbers. It makes more sense when you express scales as fractions. 1/10,000 is larger than 1/500,000, which is larger than 1/1,000,000.

We can easily imagine scales on paper maps, although the rule is the same as on the map canvas. On a 1:250,000 map, one centimeter on our computer screens equals 2500 meters in the real world. To calculate this value, GIS software use the DPI (dots per inch) value of our screens to produce accurate ratios.

By using scales instead of fixed zoom levels, GIS software offers a great amount of flexibility. For example, we can arbitrarily change the **Scale** value on our status bar and QGIS automatically jumps to that given scale. The definition of map scale will follow us along during our work as QGIS (like most of the GIS software) uses scale definitions in every zoom-related problem. Let's see one of them--the scale dependent display. We can set the minimum and maximum scales for every layer and QGIS won't render them out of those bounds. Let's right-click on one of the layers and select **Properties**. Under the **General** tab, we can check **Scale dependent visibility**. After that, we can provide bounds to that layer. By providing a minimum value of **1:500,000** to the layer and leaving the maximum value unbounded (**0**), we can see the layer disappearing on 1:500,001 and smaller scales:

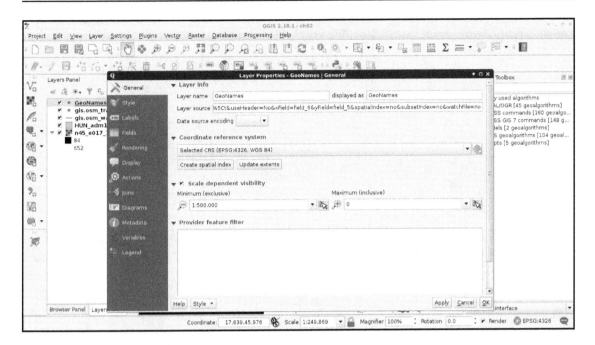

Don't bother with the scale changing by panning the map at this point. We will discuss that later, along with projections.

Summary

In this chapter, we acquainted ourselves with the GUI of QGIS and explained about data models in GIS. With this knowledge, it will be easier to come up with specific workflows later as we have an idea how the input data work and what we can do with it. We also learned about one of the mandatory cartographic elements--the map scale.

In the next chapter, we will use our vector data in another way. We will make queries on our vector layers by using both their attribute data and geometries. Finally, we will learn how to join the attributes of different layers in order to make richer layers and give ourselves more options on further visualization and analysis tasks.

3
Using Vector Data Effectively

In the previous chapter, we learned how vector data compares to raster data. Although every feature can only represent one coherent entity, it is a way more powerful and flexible data model. With vectors, we can store a tremendous amount of attributes linked to an arbitrary number of features. There are some limitations but only with some data exchange formats. By using spatial databases, our limitations are completely gone. If you've worked on a study area with rich data, you might have already observed that QGIS has a hard time rendering the four vector layers for their entire extent. As we can store (and often use) much more data than we need for our workflow, we must be able to select our features of interest.

Sometimes, the problem is the complete opposite--we don't have enough data. We have features which lack just the attributes we need to accomplish our work. However, we can find other datasets with the required information, possibly in a less useful format. In those cases, we need to be able to join the attributes of the two layers, giving the correct attributes to the correct geometry types.

In this chapter, we will cover the following topics:

- Querying and filtering vector layers
- Modifying the attribute table
- Joining attributes

Using the attribute table

The first task in every work is to get used to the acquired data. We should investigate what kind of data it holds and what can we work with. We should formulate the most fundamental questions for successful work.

Is there enough information for my analysis? Is it of the right type and format? Are there any No-Data values I should handle? If I need additional information, can I calculate them from the existing attributes? Some of these questions can be answered by looking at the attribute table, while some of them (especially when working with large vector data) can be answered by asking QGIS. To ask QGIS about vector layers, we have to use a specific language called SQL or Structured Query Language.

SQL in GIS

SQL is the query language of relational databases. Traditionally, it was developed to help make easy and powerful queries on relational tables. As attribute data can be considered tabular, its power for creating intuitive queries on vector layers is unquestionable. No matter if some modern GIS software uses an object-oriented structure for working with vector data internally, the tradition of using SQL, or near-SQL syntax for creating simple queries has survived. This language is very simple at its core, thus, it is easy to learn and understand. A simple query on a database looks like the following:

```
SELECT * FROM table WHERE column = value;
```

It uses the * wildcard for selecting everything from the table named `table`, where the content of the column named `column` matches `value`. In GIS software, like QGIS, this line can be translated to the following:

```
SELECT * FROM layer WHERE column = value;
```

Furthermore, as basic queries only allow selecting from one layer at a time, the query can be simplified as the software knows exactly which layer we would like to query. Therefore, this simple query can be formulated in a GIS as follows:

```
column = value
```

There is one final thing we have to keep in mind. Based on the software we use, these queries can be turned into real SQL queries used on internal relational tables, or parsed into something entirely different. As a result, GIS software can have their own SQL flavor with their corresponding syntactical conventions. In QGIS, the most important convention is how we differentiate column names from regular strings. Column names are enclosed in double quotation marks, while strings are enclosed in single ones. If we turn the previous query into a QGIS SQL syntax and consider `value` as a string, we get the following query:

```
"column" = 'value'
```

Selecting features in QGIS

There are usually two kinds of selection methods in GIS software. There is one which highlights the selected features, making them visually distinguishable from an other one. Selected features by this **soft** selection method may or may not be the only candidates for further operations based on our choice. However, there is usually a **hard** selection called filtering. The difference is that the filtered-out features do not appear either on the canvas or in the attribute table. QGIS makes sure to exclude the filtered-out features from every further operation like they weren't there in the first place. There is one important difference between the style of selection and filtering--we can select features with the mouse; however, we can only filter with SQL expressions.

First, let's select a single feature with the mouse. To select features in QGIS, we have to select the vector layer containing the feature in the **Layers Panel**. Let's select the administrative boundaries layer then click on the **Select Features by area or single click** tool. Now we can select our study area by clicking it on the map canvas, as shown in this screenshot:

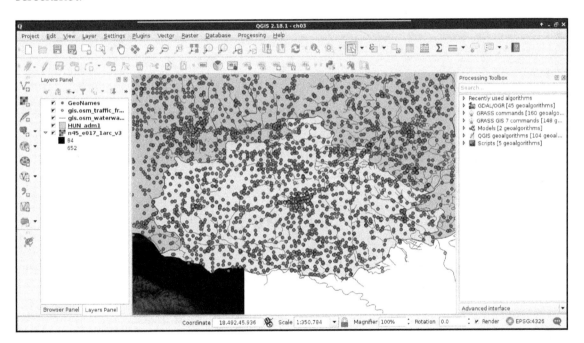

We can see the feature highlighted on the map and in the attribute table. If we open the attribute table of the layer, we can see our selected one distinguished as we have only a limited number of features in the current layer. If there are a lot of features in a layer, inspecting the selected features in the attribute table becomes harder. To solve this problem, QGIS offers an option to only show the selected features. To enable it, click on the **Show All Features** filter and select **Show Selected Features**:

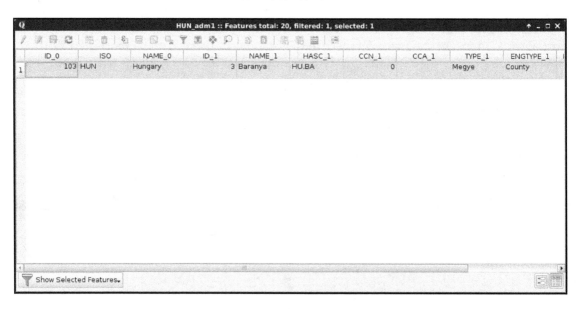

 Don't forget to try out the other selection tools by clicking on the arrow next to the current one. You can finish drawing a polygon by right-clicking on the canvas.

The attribute table and the map canvas are interlinked in QGIS. If you click on the row number of a feature, it becomes selected and therefore, highlighted on the canvas.

Preparing our data

From the currently opened vector layers, the GeoNames layer has the largest attribute table with the most kinds of attributes. However, as the extract does not contain headers, it is quite hard to work with it. Fortunately, CSV files can be edited as regular text files or as spreadsheets. As the first step, let's open the GeoNames file with a text editor and prepend a header line to it. It is tab delimited; therefore, we need to separate the field names with tabs. The field names can be read out from the readme file in order.

In the end, we should have a first line looking something like this:

```
geonameid name asciiname alternatenames latitude longitude
    featureclass
featurecode countrycode cc2 admin1 admin2 admin3 admin4 population
    elevation
dem timezone modification
```

If you use a text or code editor which replaces tabs with spaces, don't forget to switch off that feature before adding the header line. Furthermore, copy-pasting the preceding code block will probably not work.

Now we can remove our GeoNames layer from the layer tree and add it again. In the form, we have to check the option **First record has field names**. If we do so, and name the **latitude** and **longitude** fields accordingly, we can see QGIS automatically filling the *X* and *Y* fields:

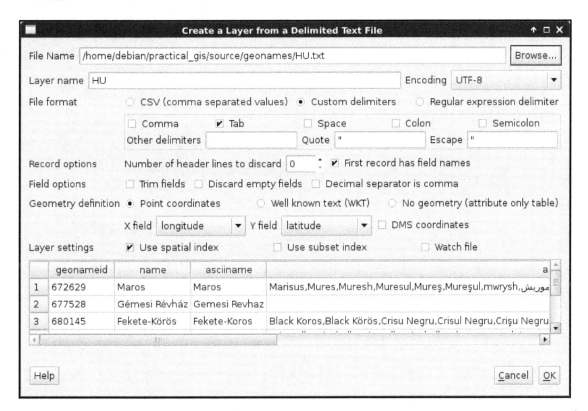

You can speed up rendering and spatial querying by checking **Use spatial index**. It will consume some memory though, so be careful with unusually large files. If you have such a big GeoNames layer that you cannot work with (like the whole U.S. table on a weaker computer), you can choose another layer for the next part.

Writing basic queries

Let's select the modified GeoNames layer and open the **Select features using an expression** tool. We can see QGIS's expression builder, which offers a very convenient GUI with a lot of functions in the middle panel, and a small and handy description for the selected function in the left panel. We do not even have to type anything to use some of the basic queries as QGIS lists every field we can access under the **Fields and Values** menu. Furthermore, QGIS can also list all the unique values or just a small sample from a column by selecting it and pressing the appropriate button in the left panel:

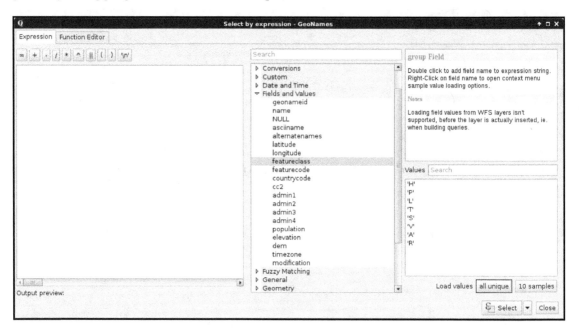

If you are familiar with basic SQL syntax, you can run some queries accommodating yourself with QGIS's query dialog and continue with filtering layers.

The basic SQL expressions that we can use are listed under the **Operators** menu. There, every operator is a valid PostgreSQL operator, most of them are commonly found in various GIS software. Let's start with some numeric comparisons. For that, we have to choose a numeric column. We can use the attribute table for that.

> TIP
> You can check the attribute types by right-clicking on the layer, selecting **Properties**, and navigating to the **Fields** tab. However, it is easy to distinguish between numeric and text fields from the attribute table. Numbers are aligned to the right in the attribute table, while strings are aligned to the left.

For basic numeric comparisons, let's choose the `population` column. In this first query, we would like to select every place where the population exceeds `10000` people. To get the result, we have to supply the following query:

```
"population" > 10000
```

We can now see the resulting features as on the following screenshot:

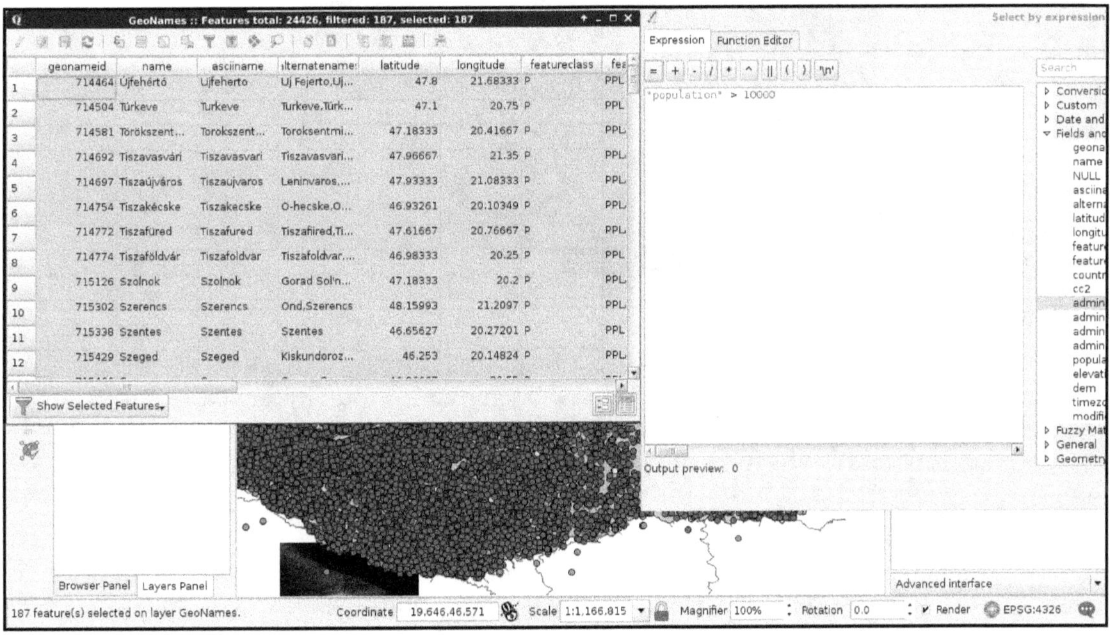

If we would like to invert the query, we have an easy task, which is as follows:

```
"population" <= 10000
```

We could do this as population is a graduated value. It changes from place to place. But what happens when we work with categories represented as numbers? In this next query, we select every place belonging to the same administrative area. Let's choose an existing number in the **admin1** column, and select them:

```
"admin1" = 10
```

The corresponding features are now selected on the map canvas:

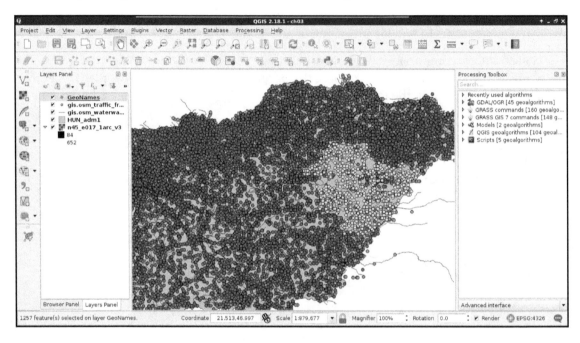

The canvas in the preceding screenshot looks beautiful! But how can we invert this query? If you know about programming, then you must be thinking about linking two queries logically together. It would be a correct solution; however, we can use a specific operator for these kinds of tasks, which is as follows:

```
"admin1" <> 10
```

The <> operator selects everything which is not equal to the supplied value. The next attribute type that we should be able to handle is string. With strings, we usually use two kinds of operations--equality checking and pattern matching. According to the GeoNames readme, the featurecode column contains type categories in the character format. Let's choose every point representing the first administrative division (ADM1), as follows:

```
"featurecode" = 'ADM1'
```

Of course, the inverse of this query is exactly the same like in the previous query (<> operator).

 We can also use relational operators on strings. If we do so, QGIS treats strings as tuples of character codes and compares them one by one. For example, if we supply the query `"featurecode" < 'AREA'`, QGIS selects everything starting with ADM, hence, *A (character code 65_{10}) = A (character code 65_{10}), but D (character code 68_{10}) < R (character code 82_{10})*, therefore, it doesn't have to search further.

As the next task, we would like to select every feature which represents some kind of administrative division. We don't know how many divisions are there in our layer and we wouldn't like to find out manually. What we know from `http://www.geonames.org/export/codes.html` is that every feature representing a non-historic administrative boundary is coded with ADM followed by a number. In our case, pattern matching comes to the rescue. We can formulate the query as follows:

```
"featurecode" LIKE 'ADM_'
```

In pattern matching, we use the `LIKE` operator instead of checking for equality, telling the query processor that we supplied a pattern as a value. In the pattern, we used the wildcard _, which represents exactly one character. Inverting this query is also irregular as we can negate `LIKE` with the `NOT` operator, as follows:

```
"featurecode" NOT LIKE 'ADM_'
```

Now let's expand this query to historical divisions. As we can see among the GeoNames codes, we could use two underscores. However, there is an even shorter solution--the % wildcard. It represents any number of characters. That is, it returns true for zero, one, two, or two billion characters if they fit into the pattern:

```
"featurecode" LIKE 'ADM%'
```

A better example would be to search among the alternate names column. There are a lot of names for every feature in a lot of languages. In the following query, I'm searching for a city named Pécs, which is called Pecs in English:

```
"alternatenames" LIKE '%Pecs%'
```

The preceding query returns the feature representing this city along with 11 other features, as there are more places containing its name (for example, neighboring settlements).

As I know it is called Fünfkirchen in German, I can narrow down the search with the AND logical operator like this:

```
"alternatenames" LIKE '%Pecs%' AND "alternatenames" LIKE
 '%Fünfkirchen%'
```

The two substrings can be anywhere in the alternate names column, but only those features get selected whose record contains both of the names. With this query, only one result remains--Pécs. We can use two logical operators to interlink different queries. With the AND operator, we look for the intersection of the two queries, while with the OR operator, we look for their union. If we would like to list counties with a population higher than 500000, we can run the following query:

```
"featurecode" = 'ADM1' AND "population" > 500000
```

On the other hand, if we would like to list every county along with every place with a population higher than 500000, we have to run the following query:

```
"featurecode" = 'ADM1' OR "population" > 500000
```

The last thing we should learn is how to handle null values. Nulls are special values, which are only present in a table if there is a missing value. It is not the same as 0, or an empty string. We can check for null values with the IS operator. If we would like to select every feature with a missing admin1 value, we can run the following query:

```
"admin1" IS NULL
```

Inverting this query is similar to pattern matching; we can negate IS with the NOT operator as follows:

```
"admin1" IS NOT NULL
```

Filtering layers

The filtering dialogue can be accessed by right-clicking on a layer in the **Layers Panel** and selecting **Filter**. As we can see in the dialogue, filtering expressions are much more restrictive in QGIS as they only allow us to write basic SQL queries with the fields of the layer. Let's inspect our study area in the administrative boundaries layer with the **Identify Features** tool, select a unique value like its name, and create a query selecting it. For me, the query looks like the following:

```
"NAME_1" = 'Baranya'
```

Applying the filter removes every feature from the canvas other than our study area:

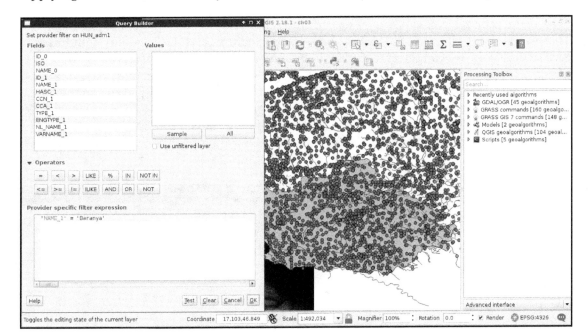

Now the only feature showing up on the canvas is our study area. If we look at the layer's attribute table, we can only see that feature. Now every operation is executed only on that feature. What we cannot accomplish with filtering is increasing the performance of subsequent queries and analyses. Rendering performance might be increased, but, for example, opening the attribute table requires QGIS to iterate through every feature and fill the table only with the filtered ones.

Let's practice filtering a little more by creating a filter for the GeoNames layer, selecting only points which represent first-level administrative boundaries. To do this, we have to supply the following query:

```
"featurecode" = 'ADM1'
```

If you have a very large GeoNames table, don't apply the filter at this point. It will be enough if you apply it after we extract a subset you can work with.

Spatial querying

We can not only select features by their attributes, but also by their spatial relationships. These queries are called spatial queries, or selecting features by their location. With this type of querying, we can select features intersecting or touching other features in other layers. The most convenient mode of spatial querying allows us to consider two layers at a time, and select features from one layer with respect to the locations of features in the other one. First of all, let's remove the filter from our GeoNames layer. Next, to access the spatial query tool in QGIS, we have to browse our **Processing Toolbox**. From **QGIS geoalgorithms**, we have to access **Vector selection tools** and open the **Select by location** tool:

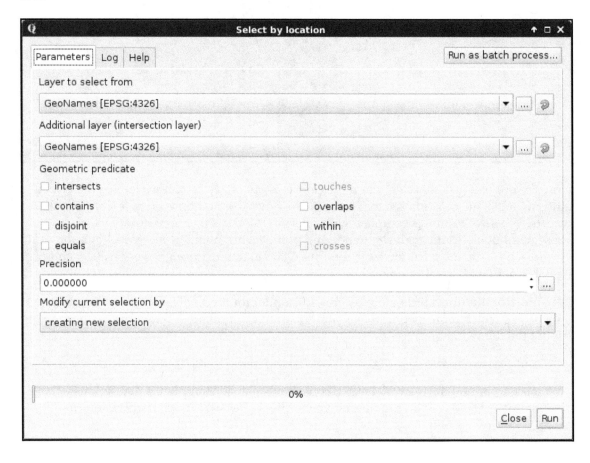

If you have a painfully large GeoNames table, select the **Extract by attributes** tool instead. The two tools have almost the same dialog, you only have to select a path to your output file, which will contain the selected features only.

As we can see, QGIS offers us a lot of spatial predicates (relationship types) to choose from. Some of them are disabled as they do not make any sense in the current context (between two point layers). If we select other layers, we can see the disabled predicates changing. Let's discuss shortly what some of these predicates mean. In the following examples, we have a layer A from which we want to select features, and a layer B containing features we would like to compare our layer A to:

- **Intersects**: Selects every feature from A which intersects any feature in B.
- **Contains**: Selects every feature from A which contains (fully encapsulate) any feature in B.
- **Disjoint**: Selects every feature from A which does not intersect any feature in B (inverse of intersects).
- **Equals**: Selects every feature from A which can be also found in layer B (can be used to check for duplicates, that is, if A and B are the same).
- **Touches**: Selects every feature from A whose boundary intersects the boundary of any feature in B, but not its interior. The interior and boundary of a point are the same, while in a line string, the boundary consists of the two nodes (end points) and the interior is everything between them.
- **Within**: Selects every feature from A which is contained (fully encapsulated) by any feature in B.

These examples are not complete explanations of the spatial predicates, but more of an attempt to give you a feel of how they work. Furthermore, there are two more predicates (overlap and cross) that we do not discuss here as they would require more theory. If you are interested, you can look at the nice Wikipedia article **DE-9IM (Dimensionally Extended 9 Intersection Model)** at `https://en.wikipedia.org/wiki/DE-9IM`.

Let's select every feature from the GeoNames layer in our study area. As we have our study area filtered, we can safely pass the administrative layer as the **Additional layer** parameter. The only thing left to consider is the spatial predicate. Which one should we choose? You must be thinking about **intersects** or **within**. In our case, there is a fat chance that both of them yield to the same result. However, the correct one is **intersects**, as **within** does not consider points on the boundary of the polygon. After running the algorithm, we should see every point selected in our study area. Consider the following screenshot:

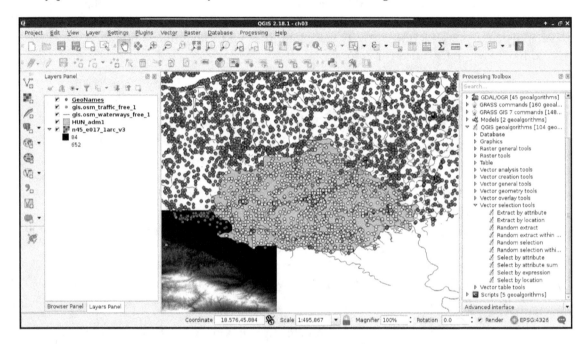

Lovely! The only problem, which I forgot to mention, is that we are only interested in features with a **population** value. The naive way to resolve this issue is to remove the selection, apply a filter on the GeoNames layer, and run the **Select by location** algorithm again. We can do better than that. If we open the query builder dialog, we can see some additional options next to **Select** by clicking on the arrow icon. We can add to the current selection, remove from it, and even select within the selection. For me, that is the most intuitive solution for this case. We just have to come up with a basic query and click on **Select within selection**:

```
"population" > 0
```

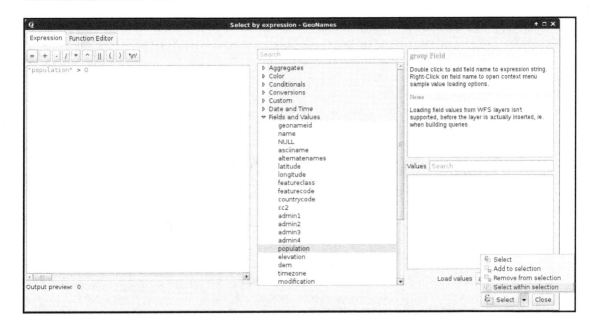

You can also invert the query (`"population"` = 0) and use the **Remove from selection** option, if it fits you better.

Writing advanced queries

We discussed earlier that basic SQL expressions in GIS allow us to select features in only one layer. However, QGIS offers us numerous advanced options for querying even between layers. Unlike filtering, the query dialog allows us to use this extended functionality. These operations are functions which require some arguments as input and return values as output. As we can deal with many kinds of return values, let's discuss how queries work in QGIS.

First, we build an expression. QGIS runs the expression on every feature in the queried layer. If the expression returns true for the given feature, it becomes selected or processed. We can test this behavior by opening the query dialog and simply typing TRUE. Every feature gets selected in the layer as when this static value is evaluated, it yields to true for every feature. Following this analogy, if we type FALSE, none of the features get selected. What happens when we get a non-Boolean type as a return value? Well, it depends. If we get 0 or an empty string for a feature, it gets excluded, while if it is evaluated to another number, or a string, it gets selected. If we get an object as a result, that too counts as false.

If we use the advanced functionality of the query builder, we get access to numerous variables besides functions. Some of these variables, starting with $, represent something from the current feature. For example, $geometry represents the geometry of the processed feature, while $area represents the area of the geometry. Others, starting with @, store global values. Under the **Variables** menu, we can find a lot of these variables. Although they do not show the @ character in the middle panel, they will if we double click on them.

Let's create a query which does the same as our last one. We have to select every feature in our study area from the GeoNames layer which has a **population** value higher than 0. Under the **Geometry** menu in the middle panel, we can access a lot of spatial functions. The one we need in our case is **intersects**. We can see in the help panel that it requires two geometries and it returns true if the two geometries intersect. Accessing the geometries of the point features is easy as we have a variable for that. So far, our query looks like this:

```
intersects($geometry, )
```

Watch out for parentheses. When you double-click on a function, QGIS imports it with only the opening one. We have to manually add the closing parenthesis.

Here comes the tricky part. We have to access a single constant geometry from another layer. If we browse through the available functions, we can almost instantly bump into the **geometry** function, which returns a geometry of a feature:

```
intersects($geometry, geometry( ))
```

As **geometry** can only process features, the last step is to extract the correct feature from the administrative boundaries layer. Under the **Record** menu, we can see the most convenient function for this task--**get_feature**. The function requires three arguments--the name or ID of a layer, the attribute column, and an attribute value. It's just a basic query in a functional form. After passing the required arguments, our query looks similar to the following:

```
intersects($geometry, geometry(get_feature('HUN_adm1', 'NAME_1',
    'Baranya')))
```

Although NAME_1 is a column name, we have to pass it as a regular string in a function. In query builder functions, we can only pass strings, numbers, and objects in the form of variables.

Now we have a constant geometry, the geometries of the point features one by one, and a function checking for their intersections. The only thing left to supply is the population part. We can easily join that criterion with a logical operator as follows:

```
intersects($geometry, geometry(get_feature(
    'HUN_adm1', 'NAME_1', 'Baranya'))) AND "population" > 0
```

For a visual example of the query, consider the following screenshot:

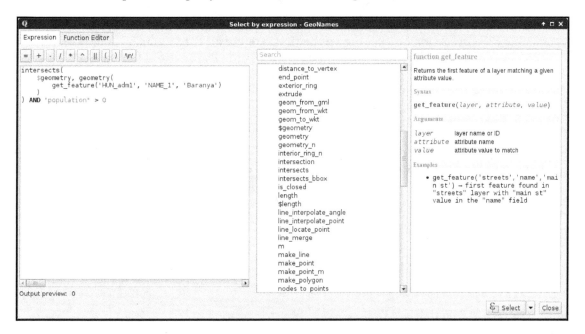

 Using the query builder instead of **Select by location** does not have a performance impact. The main benefits of using the query builder are its extended functionality and increased flexibility.

Modifying the attribute table

We can not only use the attribute tables of layers, but we can also extend, decrease, or modify them. These are very useful functions for maintaining a layer. For example, as data often comes in a general format with a lot of obsolete attributes, which is practically useless for our analysis, we can get rid of it in a matter of clicks. The size of the attribute table always has an impact on performance; therefore, it is beneficial to not store superfluous data.

Removing columns

In the first example, let's delete some values from our administrative boundaries layer. If we inspect its attribute table, we can see some unnecessary columns. In my table, ID_0 and CCN_1 have constant values, which have absolutely no meaning to me. The CCA_1 and NL_NAME_1 columns are filled with null values, while VARNAME_1 is scarcely filled, therefore, I cannot use them. Let's pick every unnecessary column and remember their names. In the attribute table's toolbar, we can see some tools related to data management. We can add and remove columns; however, those options are disabled. We can enable them by starting an edit session by clicking on the first, **Toggle editing mode** button. With the editing options enabled, we can proceed by activating the **Delete field** tool, selecting the columns we would like to remove, and approving the operation with the **OK** button:

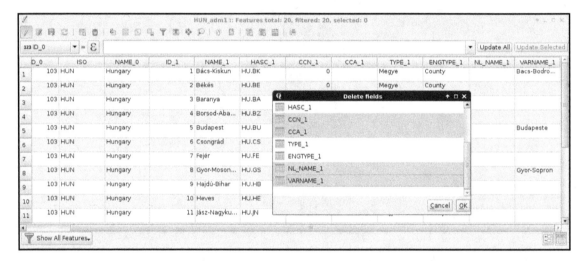

When the operation stops, we can see our unnecessary columns removed. If we save our edits, QGIS overwrites our layer and they are gone for good. Of course, we can change our mind or realize we accidentally removed some important data. In this case, we can restore the original conditions by not saving our edits.

Joining tables

While superfluous columns are often present in general data, it is not rare if we don't have the required attributes we would like to work with. If we are lucky, we can generate them based on other existing attributes, although we should not worry if this is not the case. If we can prepare a table which can be joined to the existing one on a matching column, we can easily join them together.

For this example, I prepared a small table containing descriptions of our GeoNames layer's `featureclass` and `featurecodes` columns based on the official GeoNames code page mentioned before. It is called `geonames_desc.csv` and you an access it from the supplementary material's `ch03` folder or download it directly from `https://gaborfarkas.github.io/practical_gis/ch03/geonames_desc.csv`. The formatting of this table resembles the original GeoNames table as it is tab separated; however, it does not have any geometries. It only contains two columns--a code and a description. Let's open the table with the **Add Delimited Text Layer** tool. The first line is the header and the separator is the tab character. As we have no geometries, we should also state that by checking the **No geometry** option:

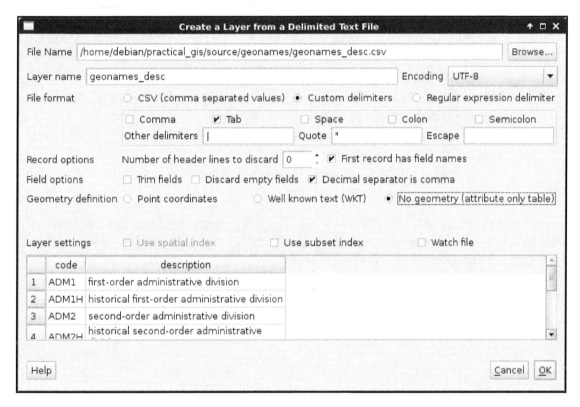

When the table is opened, we can see its entry in the **Layers Panel**. It has a special icon as it only consist of attributes. Now we can join the two layers. To start a join, we have to open the **Properties** of the target layer, in our case, GeoNames. There is a tab named **Joins**, which offers tools for managing different joins. These kinds of attribute joins do not result in overwriting the target layer, they are handled in memory; therefore, we can dynamically change them (add new ones, modify, and remove existing ones).

A successful join in QGIS needs some conditions to be met. We need a common column in both the tables as keys. These key columns hold the join conditions. The join procedure pairs these key columns together and joins the other columns of the joined layer accordingly. Therefore, to avoid ambiguities, we should have a target key column without null values and a joined key column with unique values. The key column of the joined table is never included in a join as it would introduce unnecessary redundancy. We can define a join the following way:

1. Access the **Add vector join** dialog with the green plus icon.
2. Fill the **Join layer** parameter, which is the layer or table we would like to join. In our case, it is the recent `geonames_desc` table.
3. Fill the **Join field** parameter, which is the key column of the joined layer. In our case, it is the `code` column.
4. Fill the **Target field** parameter, which is the key column of the target layer. In our case, it is the `featureclass` column.

We can also select the columns that we would like to join from the target table. As we have only two columns and one of them is the key column, we don't have to limit them. There is one final option for the prefix. As we can have an arbitrary number of joins and different tables can have the same column names, QGIS offers us the ability to prefix the target table's column names with the table's name. We can safely remove the prefix as we won't have further joins. To confirm the join, we have to click on **OK** not only in the dialog but also in the **Properties** window as simply closing it is the same as clicking on **Cancel**:

If we open the attribute table of our GeoNames layer, we can see the new **description** column appended. Furthermore, if we open the query builder, select the `featureclass` field, and query all the unique values, and do the same for the `description` field, we can see the number of unique values that match. Now let's edit the join in the **Properties** window. We can do that by selecting the join entry and clicking on the pencil icon. For the **Target field**, let's select the `featurecode` column. By inspecting the attribute table again, we can see that the values have changed and represent the description of the feature codes.

Attribute-based joins in QGIS work like left outer joins in SQL. QGIS takes every row from the target layer and matches a row from the joined table if it can. If there is no matching value, it fills the row with a null value. Every excess field is dropped from the joined table. For example, our description table contains descriptions for both feature classes and feature codes. Based on the key columns, one set of them is joined while the other is dropped.

Spatial joins

Like queries, we can also perform joins based on the location of the target and reference features. For this task, we need geometries in both the layers. This is a very handy operation when there are no common columns for joining and making one would require excessive work. To avoid ambiguity, we must have a spatial relationship able to do a one-on-one mapping between features. If not, QGIS will either pick the first matching feature or attempt to calculate statistics from the multiple target candidates.

Let's have an example, say, we would like to fill our administrative boundaries layer with population data. Our GeoNames layer has this kind of data, but they do not have a common column. If we join the whole GeoNames layer to the polygons, we would get unpredictable results. Therefore, we need to filter our GeoNames layer in such a way that only one point remains for every polygon. We can build a filter like this; we only have to select the feature codes representing our administrative divisions. In my case, the first-level administrative division contains my study area; therefore, my filter looks like this:

```
"featurecode" = 'ADM1'
```

By applying the filter, we should get one point per administrative division:

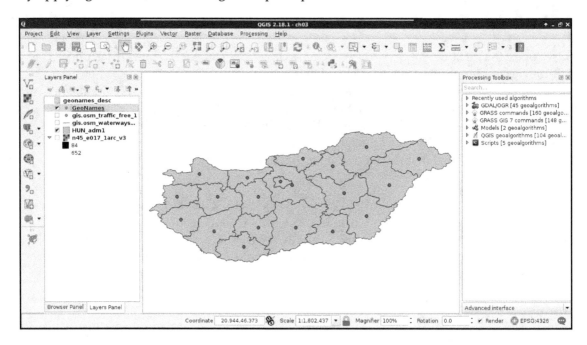

As we have a join on our GeoNames layer, QGIS asks if we would like to build a virtual layer to access the joined columns in the filter. As we don't need the description for filtering, we can say no. We will still have access to the joined column in the filtered layer.

The only thing left to do now is to run the spatial join algorithm. We can find it in the **Processing Toolbox**. Under **QGIS geoalgorithms**, we have to expand the **Vector general tools** menu, where we can find the **Join attributes by location** tool. Its dialog is similar to the **Select by location** tool, thus, we have to provide two layers and a spatial predict. The **Target vector layer** is our administrative boundary layer, the **Join vector layer** is our GeoNames layer, while the spatial predict is **intersects**. Additionally, we have to provide a join method, which should be the default, and pick the first located feature. Finally, we have to provide a path to the output file as QGIS builds a new layer with the joined tables:

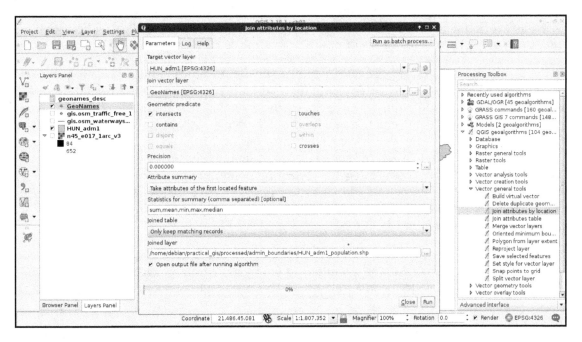

Place the new layer somewhere in your working folder, as we will need it in the next chapter.

After the algorithm finishes, we can see our new layer added to the map. The only limitation of the spatial join in QGIS is the lack of options for selecting relevant columns. It joins every column from the joined layer to our target layer. We can remove every unnecessary column from the new layer's attribute table.

Adding attribute data

There are other ways to add or modify attribute data than joining tables together. If we start an edit session, we can directly modify attribute values in the layer's attribute table. It comes in handy when we have to modify the attributes a little bit manually. If manual work is not feasible, we have a convenient tool for creating and filling new columns automatically--the field calculator. We can access this tool from the attribute table of a layer, and from the main toolbar of QGIS by clicking on the abacus icon called **Open Field Calculator**. In the field calculator, we have a dialog similar to the query builder. We have access to every function that QGIS offers and some other options related to field creation. Let's open the field calculator for our new administrative boundaries layer.

In the dialog, we can choose between creating a new field and updating an existing one. In this example, we create a new column and fill it with the population density data. As our layer now contains population data, we can easily normalize it with the area of the polygons. We would like to create a new column; therefore, we can leave the **Create a new field** option checked. After we provide a name for our new column (in my case it will be popdensity), we have to assign a type to it. As population density is a numeric value, we can choose between two types--integer and floating point number (decimal number). Even though a data value of 1.5 people sounds rather silly, we strive for accuracy and choose the decimal type. We should also set up the format of our floating point numbers. Setting the precision to 2 is enough.

The precision is deducted from the length of the field. If you assign a precision of 5 to the default of 10 field length, you will get a maximum length of 5 to the integral part and a fixed length of 5 to the fractional part.

Now we only have to provide the expression of the field. If we leave that part empty, the column still gets added but with just null values. As we can calculate the population density (people per square kilometer) from the total population and the area, which is provided in square meters, we can build the following expression:

```
"population" / ($area / 1000000)
```

Or, if we would like to get our results in people per square mile, we can provide the following one:

```
"population" / ($area / 2589988)
```

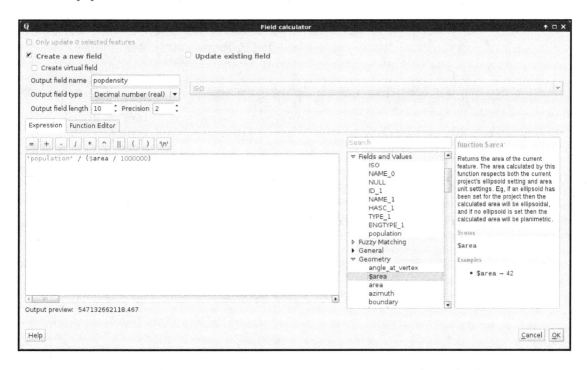

 Don't bother with that area value in the **Output preview**. The field calculator is convinced the area is in the same unit as the projection (degrees). However, those values are transformed to meters during calculations. The results still inherit the projection's distortion, but we will deal with that problem in the next chapter. You can override the default measurement unit under **Project** | **Project Properties** | **General** | **Measurements**.

Understanding data providers

While working on this chapter's examples, you might have noticed that we cannot even start an edit session on our GeoNames layer. Why do different vector data types act differently? The answer is simple and it can be found in the implementation details. Every GIS software has to decide at one point how to handle vector data.

They can build some kind of internal structure, read to and write from this structure, and handle every kind of vector data consistently during the workflow. This is one popular option. GRASS GIS does exactly this, additionally materializing this internal structure to make other processes more consistent and efficient.

QGIS, on the other hand, strives for extensibility and modularity before consistency. It utilizes one of the greatest features of object-oriented programming, polymorphism, to achieve this goal. It has a template class called `qgsVectorDataProvider` on which different data providers can implement their format-specific functionality. QGIS only communicates with this class, which can tell it the exact capabilities of the given implementation. For example, Shapefiles, among a lot of other vector formats, are handled by `qgsOgrProvider`, which can communicate with **GDAL/OGR (Geospatial Data Abstraction Library)**, a library capable of reading and writing an excessive number of raster and vector formats. This way, this provider grants the capability of editing the given layer in place; therefore, we can start an edit session in QGIS. On the other hand, `qgsDelimitedTextProvider` doesn't offer the capabilities for editing the layer in any way; therefore, we cannot start an edit session until we change our layer's format to something more capable:

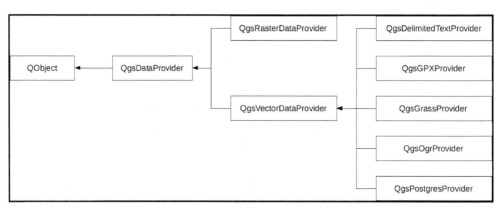

Of course, the preceding diagram only shows a few of the numerous providers that QGIS offers on the source code level. What can we do when we must edit a layer provided by an incapable vector provider? The answer is very simple--in spite of the capabilities of the given provider, once a vector layer is opened, QGIS can export it to any other supported format. Additionally, if we don't have to edit geometries, just add a few columns; QGIS has a concept for this which is called virtual fields. When we were in the field calculator, we saw an option of creating a virtual field. This is a pure QGIS concept with which we can add additional attributes on top of the existing ones in memory. Of course, we have to save the project if we would like to preserve them, or export the layer if we would like to materialize them.

Summary

In this chapter, we built on our knowledge of the vector data model and learned how to use vector data effectively with queries, filters, joins, and attribute manipulation. Great work! Now we can firmly use our vector layers to work only with the relevant part and derive more valuable information from the existing data.

In the next chapter, we will use our skills to style our data and create beautiful visualizations with it. We will also learn where, why, and how we should create a custom digital map instead of overlaying our data on already existing base maps. Additionally, we will learn how projections work and how they can aid us in our projects.

4
Creating Digital Maps

In the previous chapters, we learned about data models, and how we can use them for our needs. In this chapter, we will learn about a new GIS concept--the representation model. The representation model applies different styles and styling rules to our raw data, and creates the styled result we see on our map canvas, and later, on our digital or printed maps. By styling our maps we can decide which properties of the data are important to the readers. For example, in a thematic map showing population, we don't need the road network, while we shouldn't make a road map unnecessarily complex with population data. We can also enhance readability by adding cartographic elements to our map, like a scale bar, a navigation grid, a legend, or a north arrow.

In this chapter, we will cover the following topics:

- Styling raster and vector data
- Using different projections
- Using the print composer for creating spatial visualizations in QGIS
- Creating real maps

Styling our data

Let's start with a gentle introduction to the representation model in GIS. For these examples, we will need our modified administrative boundaries layer, our GeoNames layer, our river network, and one of our elevation maps.

 From now on, you can use the extract of your GeoNames layer, if your original one is too large.

Not only do the data models of rasters and vectors differ, but also their representation models. As rendering in every decent GIS software is hardware-accelerated, raster data are converted to textures, while vector data are tessellated in the rendering pipeline. Hence, raster values have to be mapped to 8-bit or 24-bit textures (images), while the capabilities of vector visualization depend on the implementation. The minimum capabilities are drawing icons as textures, regular shapes, connected lines, and polygons with user-defined fill and stroke colors.

Styling raster data

First, let's see our elevation model, opened in the second chapter. As we discussed before, this is the simplest rendering option that QGIS has to offer--a single-band grey representation. It simply clamps the raster values to a byte (0-255), and renders the result as an 8-bit texture. If we open the layer's properties and navigate to the **Style** tab, we can see the few options needed for such a visualization. QGIS needs a band, which is unambiguous as we have only one band, and the **Contrast enhancement** set to **Stretch to MinMax**.

Let's add some colors to this elevation model, and see how we can render it as a 24-bit image. For this, we have to change **Rendering type** to **Singleband pseudocolor**. This mode has a lot of options compared to the 8-bit mode, as it is more complex. QGIS needs to know how many colors it has to use, how to interpolate between colors, and what are the limits to the color intervals. QGIS offers a variety of predefined color ramps to choose from. As we are styling an elevation model, the **BrBG** color ramp is the best fit for our data. After choosing a color ramp, we can click on **Classify**, and QGIS automatically builds intervals for our data. As we can see, the classification results in painting the lowest points with brown, and the highest with green. We can easily invert this palette by checking in the **Invert** box. If we click on **OK**, we can see our colored elevation model:

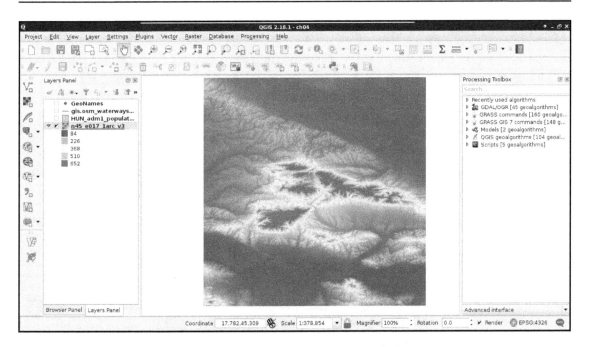

With the classification mode set to **Continuous**, we get equal intervals. The whole data range is partitioned into five equal parts, and the colors are assigned accordingly. This means, the distribution of the data are not uniform in the intervals. As my model contains values mostly between 84 and 150, I got a lot of green areas, and gradually, less brown areas.

You can see the distribution of your values under **Properties** | **Histogram**, accessed from right-clicking on a raster layer in the **Layers Panel**.

Let's change that in such a way that every interval contains the same amount of values. We can do this by changing the classification mode to **Quantile**. If we apply the changes, we can see the coloring of our model changing in a more uniform way. As QGIS does not give an aesthetic color palette for terrain visualization by default, we can import other palettes installed, but not enabled.

We can do this in the following way:

1. Click on **New color ramp** in the color chooser.
2. In the list, the **cpt-city** option contains numerous color ramps useful for geographic visualization. Select this option.
3. From the dialog's left panel, choose the **Topography** category, and import the **elevation** color ramp.
4. Give a name to the new palette.
5. Classify the data with this palette and the **Quantile** mode, and get a much more appealing result, as shown in following screenshot:

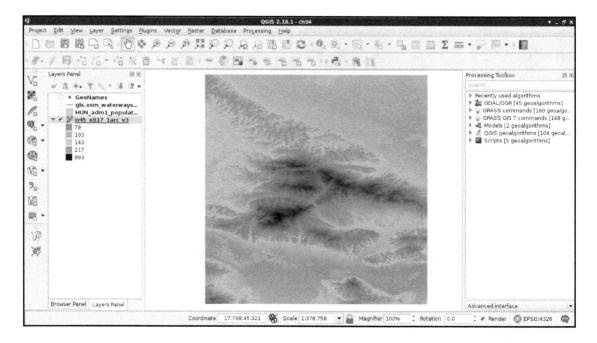

There are also other cpt-city color ramps you can download from `http://soliton.vm.bytemark.co.uk/pub/cpt-city/index.html`. To import one of the styles, download the `qgs` file, modify its extension to `xml`, and import it from **Settings** | **Style Manager** | **Import**. You can access the import button by clicking on the blue vector icon in the bottom-right corner of the window. A very fine elevation palette can be found at `http://soliton.vm.bytemark.co.uk/pub/cpt-city/td/tn/DEM_print.png.index.html`.

Let's move on to multi-band visualizations. A multi-band rendering mode needs to access three bands in the same raster. It does not matter if we have more or less bands, it just needs one band in each of the RGB channels. A very good candidate for multi-band visualization is our Landsat data. Each of the bands are 16-bit rasters (digital numbers quantized from actual reflectance data); however, they are contained in different files.

The easiest way to create a single raster from the bands is by creating a virtual raster. A virtual raster is a file that contains only references to the source rasters, therefore, it is small, but only a few software can handle it. Perform the following steps:

1. Click on **Raster | Miscellaneous | Build Virtual Raster (Catalog)**.
2. Select every band from the downloaded Landsat imagery as input files.
3. Specify a file name at a location you can easily access later. Add the `vrt` extension to the end of the file name, manually.
4. Check **Separate**, as otherwise, GDAL (as it is used by QGIS for this task) would try to merge the input rasters, and create a single-band output. This way, it keeps the input rasters in different bands.

After running the tool, our Landsat layer appears on the map canvas. We can barely see any colors in it though, as the first six bands of the Landsat 8's Operation Land Imager (its multispectral instrument) have the following spectral properties:

Band number	Name and use cases	Wavelength (µm)
1	Coastal blue (shallow waters, aerosol)	0.433-0.453
2	Blue (visible blue)	0.450-0.515
3	Green (visible green)	0.525-0.600
4	Red (visible red)	0.630-0.680
5	Near infrared (vegetation, plant health)	0.845-0.885
6	Shortwave infrared (humidity, soil type, rock type)	1.560-1.660

Therefore, in order to get a colored image, we have to create a 4-3-2 combination. To achieve this, we have to open the **Properties** of our Landsat layer, navigate to **Style**, and choose **Band 4** for **Red band**, **Band 3** for **Green band**, and **Band 2** for **Blue band**.

Now we have a colored image, although the image is quite pale and bright:

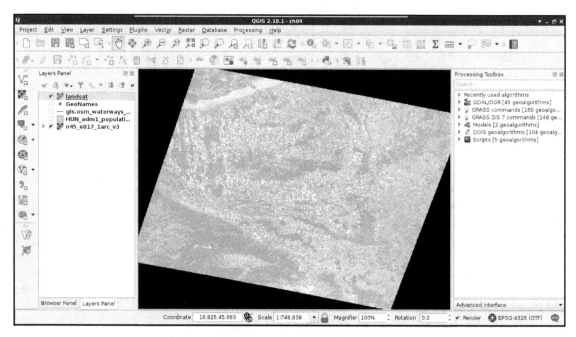

The bad news is that we have to calculate the original reflectance or radiance values, possibly with some atmospheric corrections, in order to get satellite imagery with the vivid colors that we are used to. However, we can get drastically better results even with some naive color enhancement techniques. To understand some of these techniques, let's learn why we got such a dull result. The type of the image is 16-bit unsigned integer. Therefore, it has a minimum value of 0, and a maximum value of $2^{16} - 1 = 65,535$. The visible bands (most likely due to the high reflectance of clouds) have maximum values near the absolute maximum, although the majority of their values range between 0 and 11,000.

You can observe these data in the **Histogram** and the **Metadata** tabs of the **Properties** window. In the **Metadata** tab, look for the textbox at the bottom.

When clamping values to a single byte, QGIS accepts user-defined values for minimum and maximum. If we provide values other than the minimum and maximum of our data, it truncates every value outside of this range to 0 and 255, and stretches only the in-between values. As a result, if we increase the maximum value, the values in between become less dominant, as they are stretched on a wider range.

Hence, QGIS is smart--it saw that stretching to the whole data range of our Landsat imagery is hardly beneficial, as it would produce a very dark image. Therefore, it used a technique called **cumulative cut**, and cut the outer 2% of our data in order to remove distortions caused by outliers. However, this method also discarded some important values in the upper range. This is why we got a dull image:

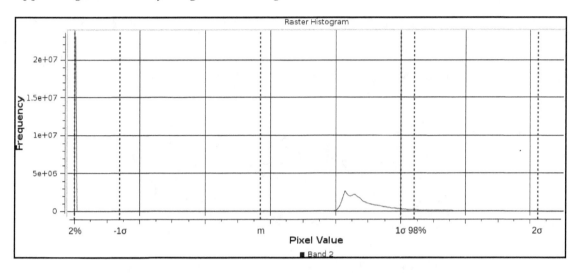

There is another popular stretching method called **σ-stretching** (**sigma-stretching**). It calculates the useful range from the mean (**m**) and the standard deviation (**σ**) of our data. The standard deviation is the density of our data in a quantified form. The more scattered our values are, the higher the standard deviation becomes, and vice versa. We can access this method by clicking on the **Load min/max values** menu in the **Style** tab. We have to check the **Mean +/- standard deviation** option, and simply click on **Load**, as **2σ** is usually a good measure for excluding outliers, while keeping the important values.

 Don't bother with the negative numbers appearing in the **Min** field. QGIS knows the type of our data is unsigned integer, therefore, replaces every negative number with 0 automatically.

If we apply our changes, we can finally see colors, although the image is still quite biased towards the upper range of the clamped values. To compensate, we can alter some values in the **Color rendering** menu of the **Style** tab. It might need a few tries to set the best values for your scene. I got a nice image with **Brightness** set to **-90**, **Saturation** set to **20**, and **Contrast** set to **10**:

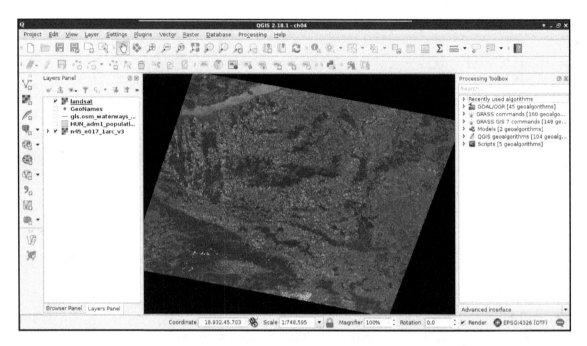

The resulting image is much more vivid, although it might be biased in one of the bands. My result, for example, has an unnatural reddish glow, which can be compensated by increasing the maximum value of the **Red band**.

Don't forget to try out other band combinations. These are called false color images, which can show properties of the land cover otherwise invisible to our eyes. For example, the 5-4-3 combination emphasizes vegetation, while the 5-6-4 combination emphasizes waters.

Styling vector data

Unlike raster data, which can be styled by their individual raster values, vector data can be styled statically, or by their attributes. Basic styling includes a simple style for a layer. For polygons, it is a simple fill with an outline, a simple line for lines, while for points, it is a scale independent circle with an outline.

If we open the **Properties** window of a vector layer, and navigate to the **Style** tab, we can see the **Single symbol** method applied to the layer.

These are cascading styles, starting in a parent class, which is predefined for the geometry type and is unchangeable. These styles are truly cascading, therefore, if we change a global attribute, like the color on any member, the whole structure conforms. A parent class (bold font) can hold multiple children. By clicking on the first (and, by default, only) child, we can customize the attributes of the style. If we choose a different styling method, and it is a complex one, it will create its own parents and children, which can be parameterized individually. By doing this, or by adding more children to the main style, we can create more and more complex styles. For example, if we choose arrows wherever possible, we can make an unaesthetic, yet interesting, visualization. Consider the following screenshot:

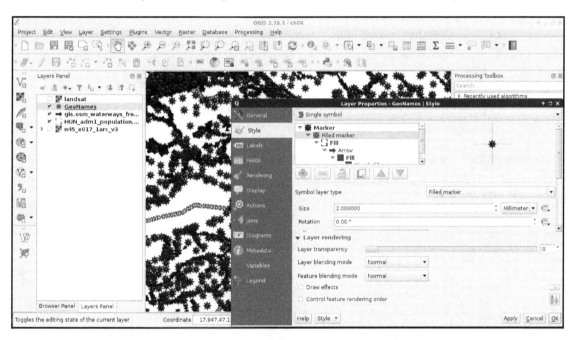

There are a lot of ways to create complex styling for a layer in QGIS. You can learn some of the numerous possibilities by trying them out. Make sure to try them out on layers with different geometry types.

These simple styles are often used to show the existence of a feature, or several features. The other common styling method is thematic mapping. With thematic styles, we can represent attributes visually. There are two distinct types of thematic styles--categorized, and graduated symbols.

Mapping with categories

Categorized symbology is useful for visualizing distinct categories on nominal and ordinal scales. This method isn't type specific, thus we can use it with strings, numbers, and any other type of attributes. For example, we can show different countries, or hotels with different ratings with different colors. To try this out, let's open the **Style** tab of our administrative layer's **Properties** window as usual. Perform the following steps:

1. Choose **Categorized** styling.
2. Choose the NAME_1 column, as it contains the names of our administrative boundaries.
3. Click on **Classify** to automatically assign random colors to distinct values.
4. Remove the last, no data entry with the minus button, as we do not have null values:

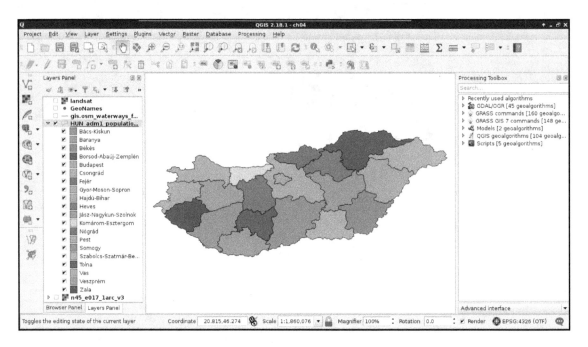

We cannot only see our boundaries colored with distinct colors, but also a legend associated with it in the **Layers Panel**. This is useful, as we can put these legends on our digital maps.

QGIS's random color generator is not a naive one. It generates appealing pastel colors to create nice representation models. However, if it is not good enough for you, you can pick great color palettes for thematic mapping from the *ColorBrewer* application. You can reach it at `http://col` `orbrewer2.org/`.

Graduated mapping

The other common method, graduated styling, is useful to show comparable attributes, effectively on interval and ratio scales. This method is type-specific, as we can only compare numbers directly. By using this method, QGIS creates intervals (bins), groups the attributes, links a color to every interval, and draws the features accordingly. We can apply graduated symbology to every geometry type, although the most common use cases are using a color ramp for shading polygons (choropleth map), and applying different icon sizes on points (proportional symbol map). First, let's apply a filter on our GeoNames layer (or its extract) to only show some of the settlements. With the following query we can filter only the seats of the administrative divisions:

```
"featurecode" LIKE 'PPLA%'
```

There is another very common type of vector symbology, called dot density. It is created by scattering points in the polygons according to a numeric column and a ratio value (for example, 1 point = 1,000 people). This is currently unavailable as a symbology type in QGIS, but it can be achieved with the **Random points inside polygons (variable)** tool found in **QGIS geoalgorithms** | **Vector creation tools**.

Now we can apply a graduated symbology by doing the following steps:

1. Open the **Style** tab of our administrative boundaries layer.
2. Select the **Graduated** symbology.
3. Select a numeric column. Population density is a nice column to work with.
4. Select a color ramp (I used **YlOrBr**).
5. Select one of the familiar modes from styling rasters. **Quantile** works really well with population density.
6. Click on **Classify**, and apply the style.
7. Open the **Style** tab of our `GeoNames` layer.
8. Select the **Graduated** symbology and the `population` column.

9. For the symbology method, select **Size**.

10. Select a mode, and click on **Classify**. With population, **Equal Interval** is a good choice for creating initial intervals and modifying them to some more appealing ranges:

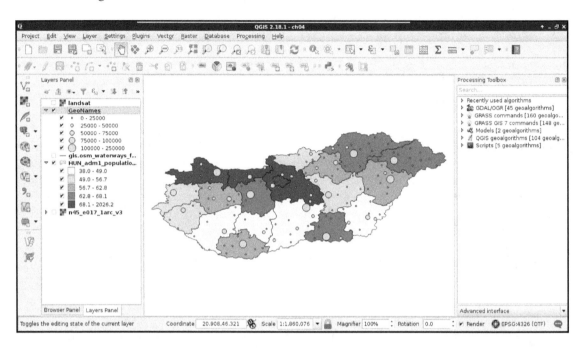

The custom ranges should always depend on the properties of the mapped data. For mapping settlements in Hungary excluding the capital city, I usually use intervals as per the preceding screenshot.

Understanding projections

As we know, spatial data can come in different projections. However, we can work with only one projection at a time. In QGIS, we can see our project's projection on the right side of our status bar. It is denoted with **EPSG:4326** for us, as this is the identifier of our projection. You might have noticed that since we added the Landsat layer, the projection changed to **EPSG:4326 (OTF)**. This change occurred as the Landsat imagery is in another projection than our project, and QGIS automatically transformed the layer with an on-the-fly (OTF) transformation.

So, if we can transform anything to a well-recognized global projection, why should we care? We can use the Mercator projection that popular web maps (like Google Maps and OpenStreetMap) use, and overlay our data on them. Well, take a look back at the first chapter where I visualized my study area, and compare the shape of Hungary with the other images. On that map, you can see most of the countries with their real sizes, and Hungary with minimal distortions. That's true--projections lie. We simply cannot map a spherical surface to a plane without distortions. In order to understand the nature of these distortions, let's see how projections work in a nutshell.

 You can see how some of the projections distort by browsing the earth application at `https://earth.nullschool.net`, choosing another projection than **Ortographic** in the settings, and panning the map.

First of all, let's get a bit technical. In GIS software, we do not use projections--we use **Coordinate Reference Systems** (**CRSs**) instead. A CRS has an ideal mathematical model of the Earth, a datum, and a projection, which maps the coordinates on the model to a flat surface.

Plate Carrée - a simple example

Let's see how the projection we used since the beginning of the book works. It is the **EPSG:4326**, or Plate Carrée (flat square), which is a coordinate system with an equirectangular projection using the **WGS84** (**World Geodesic System 1984**) ellipsoid.

We need a mathematical model to begin with, as the real shape of Earth is uneven, and therefore, very hard to represent mathematically. The most simple shape we can model Earth with is a sphere. As there is a great difference between the real size of Earth and the optimal sphere that it can be represented with, we can use a little more complex shape to increase overall accuracy--an ellipsoid. As Earth is a little flattened on the poles, an ellipsoid (technically an oblate spheroid) offers the best fit from simple shapes. It is used by most of the projections. An ellipsoid has two very important parameters--a size and flattening, which minimizes the difference from the shape of Earth for a use case. The WGS84 ellipsoid strives for the best overall accuracy.

 There is an even more complex and more accurate, but irregular, mathematical model to represent Earth, which is called the geoid. However, it requires extensive calculations to work with, therefore, it isn't used by projections. It is still used to obtain accurate height values from GPS measurements.

The second thing we need is a datum. The datum is the referenced ellipsoid. By referencing an ellipsoid, we bind its center to somewhere in or on Earth. In the Plate Carrée, we have a WGS84 datum, which binds the center of the WGS84 ellipsoid to Earth's center of mass. This is a somewhat special case, which is very easy to understand, as Earth's center of mass does not change much, only our measurements get more accurate.

The last thing we need is a projection. A projection is a method to transform coordinates on our model to a flat surface. Therefore, we need a shape which can be flattened out seamlessly, and a function to map coordinates. Plate Carrée is a perspective projection. Perspective projections work like an object put into the path of light. It casts a shadow on the surface behind it, and that shadow is the projected image of the object (in our case, the reference ellipsoid). Let's imagine our ellipsoid as a totally transparent crystal spheroid with borders of the countries painted on it as narrow black lines. We can imagine that the properties of the shadow it casts depend on two factors--the place where we put our light source (perspective point), and the shape of the surface behind it (projection surface).

Based on the perspective point, we have a lot of options, although these are three distinct, specific cases which are used often in cartography:

- **Orthographic**: The perspective point is in infinity, therefore, the light rays are parallel
- **Gnomonic**: The perspective point is in the middle of the ellipsoid; it is mainly used for polar maps
- **Stereographic**: The perspective point is at the far end of the ellipsoid

Based on the projection surface, we also distinguish between these three distinct cases:

- **Cylindrical**: We wrap our ellipsoid in a cylinder, and project the whole ellipsoid by rotating the light source. We flatten out our cylinder by cutting it, and get a rectangular map.
- **Planar**: We project our ellipsoid on a plane, and get a circle-shaped map as a result.
- **Conic**: We place a cone on the opposite side of our ellipsoid as our perspective point. We flatten out our cone by cutting it, and get a map with a shape of a half circle.

We are just scratching the topic of projections to give you an idea how they work. There are a lot more types based on how we place our surface on our model, and every type has its distinct properties and use cases. For a more in-depth, but also easy to follow guide, you can read the book *Understanding Map Projections* written by Melita Kennedy and Steve Kopp. For a less technical, informative guide, make sure you read the projection guide of Axis Maps at `https://axismaps.github.io/thematic-cartogr aphy/articles/projections.html`.

The Plate Carrée is one of the simplest perspective projections possible - it is an orthographic cylindrical projection. It is normal (the top and bottom sides of the cylinder are parallel to the Equator), thus it has a starting longitude (λ_0) of 0°, and a starting latitude (ϕ_0) of 0°. Hence, it is a simple linear mapping of geographic coordinates to the projected coordinates ($x = \lambda$, $y = \phi$):

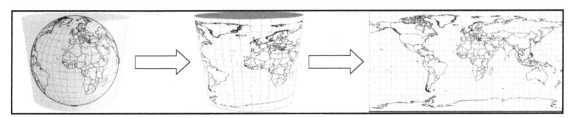

 Technically, **EPSG:4326** is the WGS84 ellipsoid (or datum) alone as a geodetic CRS. However, as every ellipsoid is mapped with a linear projection when used alone, the fact that **EPSG:4326** uses a Plate Carrée projection stands still.

Going local with NAD83 / Conus Albers

Let's see a local CRS for mapping the entire United States. NAD83 / Conus Albers (**EPSG:5072**) uses an ellipsoid called **GRS80 (Geodetic Reference System 1980)**. It has almost the same properties as WGS84, and they had the same properties back in the time. However, WGS84 underwent some changes (realizations) to give a better fit for GPS systems, and therefore, its flattening became slightly different.

More importantly, NAD83 / Conus Albers uses a local datum--**NAD83 (North American Datum 1983)**. That means, it is not referenced to Earth's center of mass, but to the North American plate. It is still referenced to the same planet, so the question arises--what's the difference? The answer is simple--plate tectonics. The coordinates on global datums are changing constantly due to plate movements. This is a very slow change (a few centimeters every year), however, it still can be an issue for high accuracy surveys and analyses. To avoid the urge for correcting old data, local datums are referenced to local places, and are moving with the plate underneath. This also means that global and local datums are slowly drifting apart. Consider the following diagram:

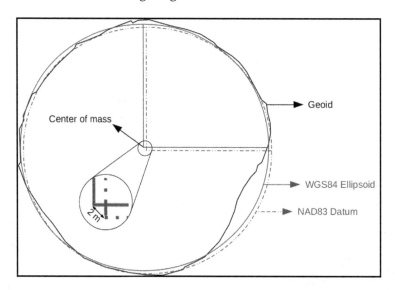

The geoid is the most accurate mathematical model we have for the Earth's shape. Its center is located in the center of Earth's mass. The WGS84 oblate spheroid is a much more general model, whose center is placed at the geoid's center. The difference between the two models is called geoid undulation, which is automatically height-corrected by GPS systems. As the NAD83 datum is referenced to the North American plate, its center slowly drifts apart from the center of the WGS84 ellipsoid.

The word datums is grammatically incorrect. Datum is the singular form of data, however, as in GIS we refer to completely different concepts with the words datum and data, I'm using the grammatically incorrect datums to avoid ambiguity.

The NAD83 / Conus Albers (EPSG:5072) uses an Albers projection. It is conic, therefore, the geographic coordinates are projected onto a cone, which, by flattening out, produces a half-circle-shaped map for its whole validity extent. However, there is a catch; it's not perspective. Therefore, the perspective point is meaningless in this concept. Non-perspective projections can only be defined mathematically, we cannot reproduce them with a light source.

If you would like to see one of the geoid models in your browser, make sure you check out the great 3D visualization at `https://www.chromeexpe` `riments.com/experiment/geoid-viewer`.

Choosing the right projection

Now that we know how projections work, let's discuss how should we choose the right projection for our work. We learned before that projections distort reality, as we cannot convert an ellipsoid to a flat surface. Projections have some properties, which are shape, area, direction, distance, scale, and bearing. From these properties, a single projection can only fully preserve a few, at the expense of other properties. Based on the preserved properties, we distinguish between the following types:

- **Conformal**: Preserves bearings, and shapes locally (still distorts shapes, but it creates the best global approximations). Conformal maps (like Mercator) were life savers back in time, when sailors only had a compass and a map. As it preserves bearing, we can connect two points with a straight line, align our compass, and walk between them based on the bearing. It distorts areas beyond recognition, though.
- **Equal-area**: Preserves areas, but distorts shapes. It is used for visualizing and analyzing data, where showing areas proportionally is important (such as using indices normalized by area). The Albers is an equal-area projection.
- **Azimuthal**: Preserves directions. Straight lines represent the shortest routes on the ellipsoid between two arbitrary points, also called great circles.
- **Equidistant**: Preserves distances from the distortion-free part or parts of the map. From its center, or another distortion-free point, we can measure a straight line, which corresponds to the real distance. This property is not preserved for other pairs of points.
- **Compromise**: Does not preserve any property, but strives for minimizing errors. Compromise projections are great for global mapping if the map doesn't have to preserve a property.

Each projection has at least one point it does not distort--its center. However, some projections have more points, in some cases, one or two lines without distortions. The Plate Carrée does not distort along the Equator, while the Albers Conic has two arbitrary distortion-free, standard parallels. NAD83 / Conus Albers has these standard parallels at ϕ_N 29.5° and ϕ_N 45.5°.

We did not talk about an important property of projections--scale. Only a few projections preserve scale, most of them distort it. However, the printed and digital maps always show a constant scale, usually with a scale bar. Also, we witnessed during our work that the scale always changes when we pan the map (Plate Carrée does not preserve scale). To overcome this issue, the scale value in these cases is an approximation based on the center of the map (or less often, some kind of average from different parts of it). It simply displays the exact scale in the center, and assumes that we know if our projection preserves or distorts scale to the edges.

Projections have another important property which we usually do not discuss in depth--the unit. Each projection is crafted in a way that distances can be measured with real-world units. Some of them use degrees, while others use SI units (most commonly, meters), feet, or miles.

Some of the CRSs do not need these kinds of considerations, as they are fitted on a small area. This means that distortions are mostly negligible in their validity extents. If we have a small enough area to map (just like our study area), we can choose such a CRS. For countries too big for an all-purpose CRS, there are multiple ones. There are CRSs to visualize the entire country with different properties, while there are also CRSs for smaller regions giving a better fit.

We do not need to know about all the existent CRSs to choose one. There are databases of CRSs which we can browse. The most widely used database is the **EPSG** (**European Petrol Survey Group**), which maintains an up-to-date catalogue of all of the popular CRSs.

These CRSs identified by their EPSG codes (such as EPSG:4326 for Plate Carrée) are supported by all kinds of GIS software, such as QGIS. Let's select a CRS from an online version of this catalogue at `http://epsg.io/`. We can type our country's name in the search field, and the site will list all of the projections for our country. We can filter our results to see only projected CRSs (exclude datums) by clicking on **Projected** on the right-hand side:

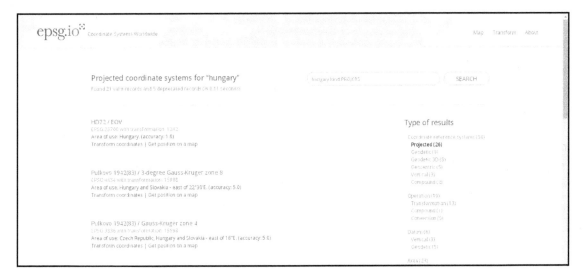

What we have to remember is the EPSG code of our preferred CRS. For example, I will work with EPSG:23700 (HD72 / EOV) from now on. In QGIS, we can change our project's projection in the following way:

1. Click on the project's current projection (**EPSG:4326**).
2. In the projection dialog, enable OTF (on-the-fly transformation) by checking in the appropriate check box.
3. In the **Filter** field, type the EPSG code from the online catalogue.

4. Select and apply the right CRS from the results:

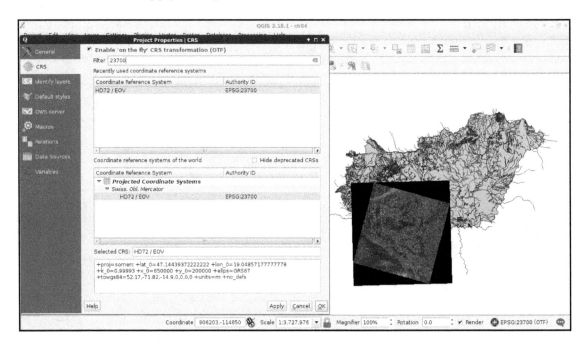

Let's see the consequences of using a more appropriate projection. If you have multiple projections for the country you are working with, choose a projection for the whole country for now. For this task, we need our administrative boundaries layer. First of all, to access the transformed metrics of our layer, we need to define an ellipsoid for measurements:

1. Open **Project** | **Project Properties** | **General**, and select the **WGS84** ellipsoid in **Measurements** | **Ellipsoid**.
2. Open the attribute table of the administrative boundaries layer, and choose the **Field Calculator** tool.
3. Name the updated population density column. I'll use the name `pd_correct`.
4. Choose the **Decimal number** as a type, and add two decimal places with the **Precision** field.
5. Calculate the column with the formula used in the last chapter (`"population"` / (`$area / 1000000`) for SI units).

If we compare the new population density column with the older one, we can see some differences. The farther our country lies from the equator, the bigger the differences are.

If you enable OTF, and select an ellipsoid in the **Measurements** | **Ellipsoid** menu, it doesn't matter what projection you are using. QGIS always returns correct values for both, area and length. Just remember--it still matters how you present your results.

Preparing a map

In this example, we will create our first real map, a road map of our study area. We will start by a hybrid map like the one we can see in Google Maps by changing to satellite imagery. For this task, we will need the roads layer from the OSM dataset (gis.osm_roads_free_1), the rivers layer (gis.osm_waterways_free_1), the water bodies (gis.osm_water_a_free_1), and the land-use layer (gis.osm_landuse_a_free_1). We will also need the GeoNames and the administrative boundaries layer. First of all, to speed up our work, let's extract only the relevant features. We should do the following steps to every vector layer except the administrative boundaries. If some of your layers are not that large, you can skip these steps for those layers:

1. Add the layer from the **Browser Panel**, or with the **Add Vector Layer** tool.
2. Open the layer's **Properties** window, go to the **General** tab, and click on the **Create spatial index** button.
3. Apply a filter on the administrative boundaries layer to only show the study area (we only have to do it once).
4. Open **QGIS geoalgorithms** | **Vector selection tools** | **Extract by location** from the **Processing Toolbox**.
5. Fill out the required fields, as we did in Chapter 3, *Using Vector Data Effectively*, and select the **Intersects** spatial predicate.
6. Choose **Save to file** by clicking on the button next to the **Extracted (location)** field, choose a destination folder, the **SHP files** format, and give a name to the result.

Be patient. QGIS geoalgorithms are much slower than their PostGIS counterparts. Despite the spatial index we created, extracting the roads layer will take a while (without spatial indexing, it would take up to half an hour for 500,000 features). Go grab a coffee (or just take your time) in the meantime.

Rule-based styling

Road data from OpenStreetMap comes with a classification, which is very useful for creating road maps. This classification is stored in the `fclass` column in our layer. If we create a categorized symbology based on that column, we can see that there are a lot of classes. From those classes, only a few are appropriate to show at this scale:

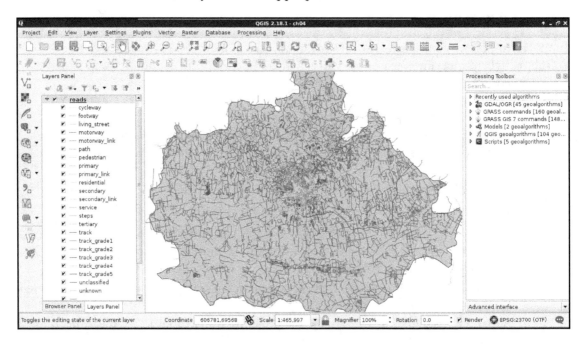

Furthermore, some of the classes belong to a single type. For example, the `motorway` and `motorway_link` classes distinguish between two subtypes of motorways. QGIS offers a great tool for these cases, called rule-based styling. Let's open the **Style** tab of our layer's **Properties** window, and choose **Rule-based**. We can remove the classification by selecting them all with the *Shift* key, and clicking on the minus button. For this scale, we will only show motorways, highways, and other important roads. We can add a rule with the plus button, which opens a dialog for creating a rule definition. By clicking on the **...** button next to the **Filter** field, we can build our first definition as follows:

```
"fclass" LIKE 'motorway%'
```

Similar to Google Maps, we create a complex line style for motorways, a thick yellow line with a thin black outline. We can do this by stacking two line styles. A 1 millimeter-wide black line goes to the bottom, while a 0.8 millimeter-wide yellow line goes on the top. This will create a 0.8 millimeter-wide yellow line with a 0.1 millimeter-wide black outline on both sides. First, we create the black line, then add a new line with the plus button. Finally, we style the new line:

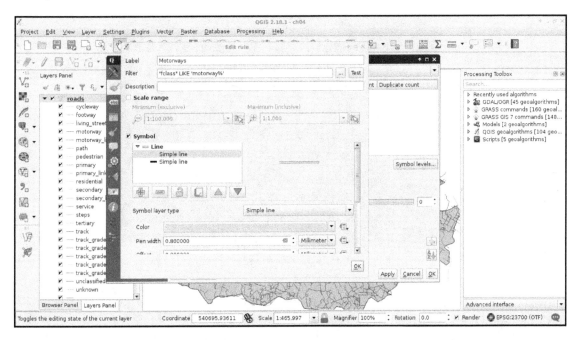

The other roads should be styled following the same analogy, just with narrower lines. For selecting the highways, we can use the following query:

```
"fclass" LIKE 'primary%'
```

I visualized highways with a 0.6 millimeter-wide yellow line on a 0.8 millimeter-wide black line. Other important roads can be selected with a similar query:

```
"fclass" LIKE 'secondary%'
```

For these roads, I created a single style, a 0.5 millimeter-wide grey line. By visualizing our roads on the top of the Landsat imagery, we can see our road map slowly getting in shape:

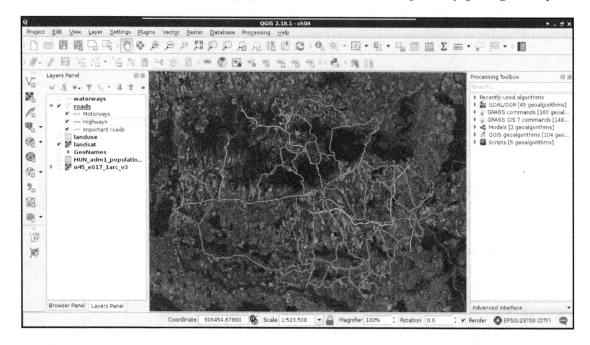

The only thing we lack is handling line connections properly. As we can see, our roads consist of several features. When those features connect, the ends of the lines are visible, and therefore, the whole image has a broken feeling. We can manipulate how those lines are rendered, though. As complex styles are rendered in different passes (layers), we can define their order. If we would like to create nice connections, we should render the black outlines first. As secondary roads are the least important, we should render them next. In the next pass, we should render highways, therefore, secondary roads connect into them directly, and only in the last pass we should render motorways. To define this order, we have to open the **Symbol levels** menu in our layer's **Style** tab. There, we just have to define the order with numbers. The higher the number, the later the style gets drawn.

If we create the ordering defined in this section, we get a more aesthetic result, as seen in the following screenshot:

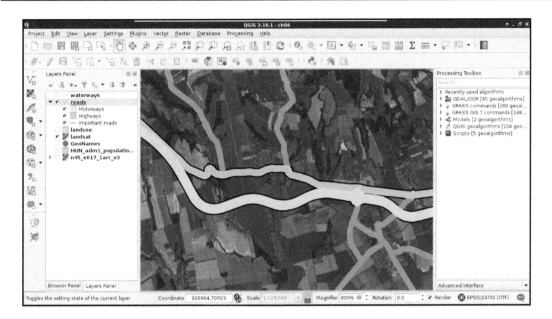

Adding labels

Now that we have some roads, we should also add some labels to them. Maps at this scale usually do not contain street names, but road numbers. For higher-degree roads, OpenStreetMaps delivers these road numbers in the **ref** column. First, let's create a simple labeling by opening the **Labels** tab in our layer's **Properties**. We can enable labeling by selecting the **Show labels for this layer** option, and choosing the **ref** column. If we apply this, we can see our road numbers on the canvas, however, it is far from an appealing map.

In the labeling window, we have a lot of options to customize our labels. Let's start with the **Text** tab. As the default text size is quite large, we can set our font size to 8 points. Also, change the color of the text to white, as it goes quite well with a colored background. The next tab we should see is the **Background**. Road labels are usually drawn in some kind of shield shape, like the tab's icon in QGIS. We can do that with a custom SVG icon, however, for the sake of simplicity, let's stick with a simple colored rectangle for now. We can enable backgrounds by checking in the **Draw background** box. Now we can customize the background with our preferred color or border style.

The next tab is an essential one for every labeling task--the **Placement**. In there, we can define if we would like to align our labels horizontally, parallel to the given feature, or curved along a line. If we choose **Horizontal**, we can only define a repeating interval.

That is only important when our features are large enough. If we click on **Apply**, we can see our labels are much better, however, they are too dense for a nice map.

> In labeling, almost every parameter has a data-defined override. If you would like to provide such a parameter using a column or an expression, you can click on the icon, which looks like a small menu next to the given parameter's field. In there, you can either choose a field, or build an expression by clicking on **Edit**.

This phenomenon occurred as our roads consist of a lot of individual features, representing some segments. QGIS automatically labels these individual features; it does not matter if some of them form logically coherent units. There are two solutions to overcome this issue. The first one involves some fiddling, and does not guarantee correct results, while the second one is more exact, but we have to run a geoalgorithm to achieve it. Let's try out the fiddling one first. We have to open the **Labels** tab of our layer's **Properties** window, navigate to **Rendering**, and check in **Merge connected lines to avoid duplicate labels**. Now QGIS tries to merge connected features with the same attribute values in memory to avoid duplicate labels. If the labeling is still too dense, we can suppress labeling of features smaller than a specified value by filling in the corresponding field:

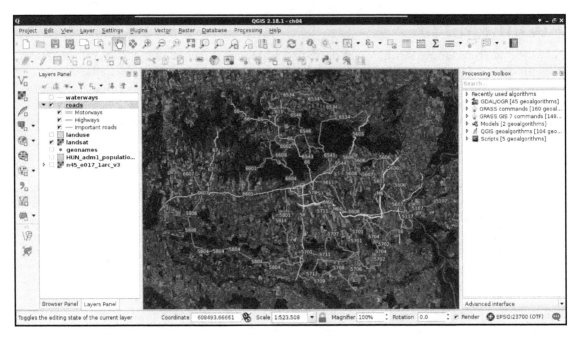

Although this method can help in creating correct labeling without the need of further geoprocessing, it did not work well in our case. The labels are still placed irregularly, some of them do not even appear.

 In QGIS 3, we will have a manual labeling tool, which can be used to manually remove, or just simply move away the ill-placed, or superfluous labels. Manual label placement in QGIS 2 is still possible, although it needs a little bit of fiddling. For more information, please read the answers at `ht tps://gis.stackexchange.com/questions/67408/how-does-manual-la bel-placement-in-qgis-1-9-work`.

The correct way to handle these cases is merging the lines in a way there will be only one feature for every unique value from the **ref** column. There is a tool for exactly this--dissolve. It needs an attribute to unify features with, and strives to merge connected features with the same attributes. If there are features which cannot be merged, it creates a multipart feature with the disconnected parts. We can dissolve our road layer in the following way:

1. Apply a filter on the roads layer, so the dissolve tool does not have to iterate through all of the features. A filter for our current visualization can be expressed with the query `"fclass" LIKE 'motorway%' OR "fclass" LIKE 'primary%' OR "fclass" LIKE 'secondary%'`.
2. Select the tool **QGIS geoalgorithms** | **Vector geometry tools** | **Dissolve** from the **Processing Toolbox**.
3. Specify the roads layer as **Input layer**, uncheck the **Dissolve all** box, and select the **ref** column as the **Unique ID field**. Also select a destination folder, and save the result to a Shapefile.

After running the algorithm, we get a dissolved layer, which only has one feature for every road number. As the attributes which were the same for the new features are retained, we can style our new layer like our previous one.

To make things even more simple, we can apply the same styling to our new roads layer in these two easy steps:

1. Right-click on the old roads layer's entry in the **Layers Panel**, choose **Styles**, and click on **Copy Style**.

2. Right-click on the new roads layer's entry, choose **Styles**, and click on **Paste Style**:

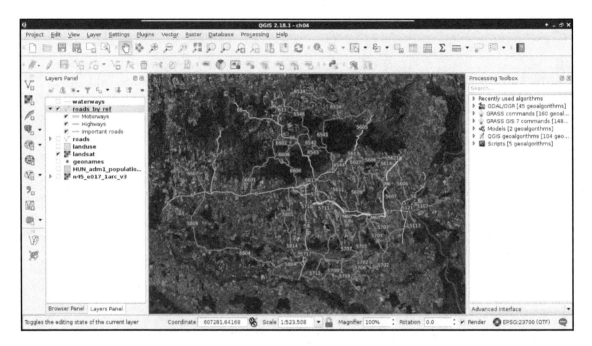

A really professional road map repeats labels (if feasible) after junctions. To create a road layer where features are split in intersections, you can use the **QGIS geoalgorithms** | **Vector overlay tools** | **Split lines with lines** tool from the **Processing Toolbox**. You have to provide the same dissolved road layer for both of the input layers.

The only thing left is adding some town labels. We will use our GeoNames layer to visualize more important settlements as follows:

1. Filter the GeoNames layer to only show more popular settlements. The seats of administrative divisions can be filtered with the query `"featurecode"` LIKE `'PPLA%'`. Note that if you created an extract in a Shapefile, the column name is truncated to `featurecod`.

2. In the **Style** tab, select the **No symbols** option.

3. In the **Labels** tab, choose **Show labels**, and select the name column for labeling.

4. In the **Text** menu, select white for text color.

5. In the **Buffer** menu, check **Draw text buffer**, specify an appropriate buffer size, and select black for its color.

Now we can also see some of the settlements labeled on our map. If some of the labels got suppressed, we can modify the weights of the GeoNames and the roads layers in **Labels** | **Rendering** | **Obstacles**.

Creating additional thematics

The core of a thematic map is surprisingly always its thematics. We can classify our maps based on the most important, most emphasized thematic (for example, we will end up with a road map), but it does not exclude adding more thematics for various cases. We can fill our map if it is too empty, or help the readers by adding more context. In this example, we replace our Landsat imagery by some thematics from the OpenStreetMap dataset. We will visualize land use types, rivers, and water bodies on our map. As we went through styling vector layers quite thoroughly before, we will only discuss the main guidelines to achieve nice results.

First of all, let's disable our Landsat layer, and enable the layers mentioned before. The first layer we will style is land use, as it will give the most context to the map. If we apply a categorized styling on that layer based on the fclass column, we can see similar results in the roads layer. There are many classes, most of them containing details, which are superfluous for this scale. To get rid of the unnecessary parts, and focus only on the important land use types, let's apply a rule-based styling:

- **Forest**: Only the **forest** category with a dark green color
- **Agriculture and grassland**: The **farm**, **grass**, **meadow**, **vineyard**, and **allotments** categories can go here with a light green color

- **Residential**: The **residential** category, visualized with a light orange color
- **Industrial**: The **industrial** and **quarry** categories, with a light grey color

As the default black outline would draw too much attention and distract readers, let's apply a 0 mm, **No Pen** outline style to every category. We can access the outline preferences by clicking on the **Simple Fill** child style element.

> If you have additional categories taking up large map space, you can create additional rules, or just fit them into the most appropriate one from the aforementioned.

Now we will do a very cool thing. The water bodies layer not only stores lakes and other still water, but also larger rivers in a polygon format. We will not only show these lakes and rivers along with the linear river features, but also label them, making the labels run in the polygons of larger rivers. For this, let's place the river layer on top of the water bodies, and give them the same light blue color.

Next, navigate to the **Labels** tab of our rivers layer, and select **Rule-based labeling**. We apply only one rule, which only labels rivers, as we wouldn't like to load the map with labels of smaller waterways. We can build such an expression similarly to the other OSM layers as follows:

```
"fclass" = 'river'
```

In the **Placement** menu, we define the labels to run on a **Curved** path along the linear features. The only allowed position should be **On line**, therefore, we uncheck **Above line** after checking it in. Finally, in the **Text** menu, we specify the text color to be a darker shade of blue in order to make it go nicely with the rivers' color. I also specified another font, which enabled a **Bold** typeset.

As the linear features of the rivers run exactly in the middle of the outlines (in the streamline), our labels are run exactly in the middle of the polygons, along the streamlines, which looks really nice and professional:

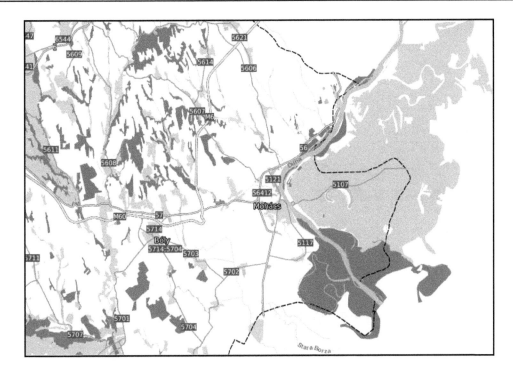

It can give a nice touch to the map if we delimit our study area. We can do so by applying an **Outline: Simple line** style instead of the **Simple Fill** one. I left that one with a black color, but increased the line width, and applied a dashed **Pen Style** to it. You can get the same result by leaving the **Simple Fill** style, and specifying **No Brush** for **Fill style**.

It's time to add one final piece to our map--topography. For this task, we will need our elevation layers. If we open them all at once, we can see that they have a little overlap, making some linear artifacts. To get rid of these disturbing lines, and to get a more manageable elevation layer at once, we first create a virtual raster from the elevation datasets. Similar to the Landsat imagery, we open **Raster** | **Miscellaneous** | **Build Virtual Raster** from the menu bar. We browse and select every SRTM raster, then select a destination file. We do not have to check anything else, as we would like to create a seamless mosaic from the input rasters, not store them in different bands. We just select a destination folder, and name our new layer. Don't forget to append the **vrt** extension manually to the file name.

The next step is to pull the new `srtm` layer strictly above the `landuse` layer. We are going to style it in a new way to show elevation with shading. Let's open its **Style** menu, and select the **Hillshade** option. It will create a shaded relief based on the provided altitude and azimuth values. Those values determine the Sun's location relative to the surface. The default values are generally good for a simple visualization. Finally, let's alter the **Blending mode** parameter. This parameter defines how our layer are blended with the layers underneath. If we choose **Overlay**, it blends the shading to the colored parts of our `landuse` layer, and leaves the white parts. We can also reduce the layer's transparency in the **Transparency** tab to make the colors less vibrant, and more like the original values:

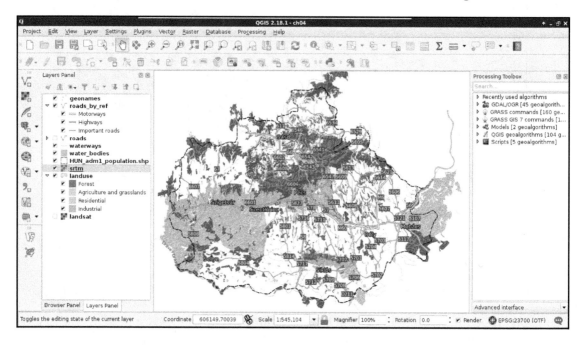

The order of the layers is important. Blending only applies to the layer it is defined on, and only takes the layers which are underneath into account. That is why we should make the `landuse` layer our bottom layer in this case, put the `srtm` strictly on its top, and put everything else on top of them. This way, the topography won't blend into the other thematics. Don't hesitate to try out the other blending modes and find out what they can do!

Creating a map

Splendid work! We managed to make the most important part of a real map--the content, or data frame. However, there are some more cartographic elements to add if we would like to call our composition a map instead of a spatial visualization. Some of the usual cartographic elements are the following:

- **Data frame**: It contains the main content of the map.
- **Title**: A short, concise title summarizing the main thematics of the map.
- **Scale and scale bar**: The scale of the map visualized with a scale bar and with a ratio number.
- **Legend**: A graphic description for the thematics of the map. It is not mandatory when the map is not thematic. For example, a shaded relief (especially a hand-drawn greyscale one) does not need a legend.
- **Attribution**: A list of sources used by the map followed by the name of the author, the copyright terms, if applied, and, at least, the year of creation.
- **North arrow**: An arrow pointing to the North if the map is not oriented that way.
- **Frame**: A small frame bounding the map, usually used with a grid and showing its reference numbers. It can show any additional information though.
- **Grid**: Either a local grid partitioning the map to logical units for easier navigation, or referencing, or a grid showing x and y axes in predefined intervals.
- **Additional data frames**: An overview map showing the mapped area in a larger context, or an inset map showing a small, but important, part of the map in greater details.

Let's add some of these elements by using the Print Composer of QGIS. To open a new composer, we can use the **New Print Composer** button on the main toolbar. We can open as many composers as we want in QGIS. If we give them talkative names, it will be quite easy to navigate between them with the **Composer Manager** tool next to the **New Print Composer** button. Although the composer is opened in a separate window, we don't have to worry about losing the composition on closing QGIS. Composers are saved with the project.

Adding cartographic elements

In the composer window, we can access the layout properties of our map instantly. In the right panel, we can choose the paper size, and its orientation. On the left toolbar, we have access to the most important cartographic elements, which are added on demand as separate, configurable items. The first four tools are for item management. We can use the **Select/Move item** tool to move and resize items, and the **Move item content** tool to pan the map inside the data frame item. Under those tools, we can access the items which can be added.

First, let's add the dataframe to the map. To do this, click on the **Add new map tool**, and draw a rectangle on the canvas. Let's resize the map to match the paper's dimensions. Once an item is added, we can snap its borders to the borders of our canvas. Now we have access to the item's properties in the right panel. Under **Item properties**, we can see the parameters of our map content. The first thing to change is its scale. It has a different scale from the browser's canvas, as it is now calculated to match our paper's size. Let's modify it to a nice, round number. When we change the scale, QGIS automatically updates the map on our canvas. However, it does not render the map at panning or zooming unless we change the **Cache** property to **Render**. Changing it degrades performance, but updates the map at every change. When you've found the right scale to use, align the map on the canvas with the **Move item content** tool:

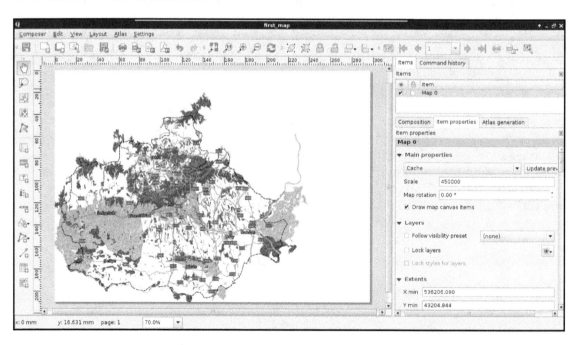

You can find the **Grids** option under the map item's properties. Add a grid with the plus sign, set the intervals in the x and y axes, and see the results. You can draw coordinates at the axes by checking the **Draw coordinates** checkbox below. Finally, change the grid's CRS to WGS84 (**EPSG:4326**), and set the intervals to somewhere between 0.2 and 0.5.

The next thing we add to the composition is a legend. We can create a legend by selecting the **Add new legend** tool, and drawing a rectangle on the canvas. As we can see, the legend is automatically created from the **Layers Panel** by default. As we have some layers which do not fit into the legend (for example, the srtm layer or the GeoNames layer), we can choose to manually customize the legend item. For this, we need to uncheck the **Auto update** box in the legend item's properties. Now we can delete superfluous entries by selecting them and clicking on the minus button. We can also rename the existing labels and groups, and change their order.

Let's get rid of the extra layers, and rename the rest of them to have more descriptive names. Also, there are two thematics (roads and land use) grouped, and two layers (waterways, administrative boundaries) ungrouped. To make the legend more consistent, let's create a custom group with the **Add group** button, name it as Other, and drag those entries into it. Finally, we should make the group fonts more consistent. The Other group has a different font, as it is a group, while the others are considered subgroups by QGIS.

To change this, you can right-click on the subgroups, and change their categorizations to group:

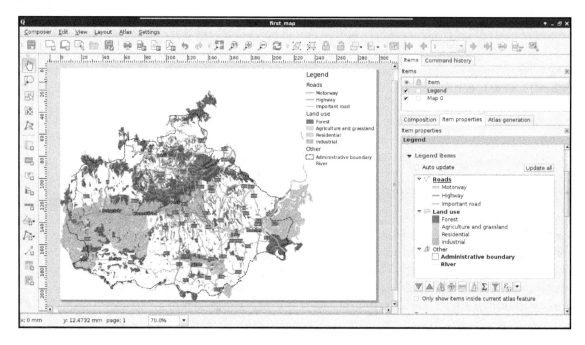

Next, we add some text content. Specifically, we add a title and proper attributions to the map. We can add custom text boxes with the **Add new label** button. Editing the label is not as interactive as in a vector editing software, but we can customize the label in its properties window in the right panel. The name of the layer should be concise, but descriptive. I used the name `My first road map - a QGIS experiment`. The attributions should go in a separate text box with a smaller font size. We should add the following four statements to the attributions:

1. OpenStreetMap data © OpenStreetMap contributors.
2. SRTM 1 Arc-second data downloaded from USGS's Earth Explorer.
3. Administrative boundaries © GADM (or Natural Earth if you used their data instead).
4. © Your Name, year of composition.

Now let's add the final piece to our map--the scale and the scale bar. We can add them both with the **Add new scalebar** tool. It comes with a fixed size, which needs some tinkering to modify. The easiest way to reduce its size is to modify the number of units it shows under **Segments | Fixed width**. We can also choose between some templates in its **Style** menu.

The second scale bar should only contain the scale in a numeric form. To achieve this, we can choose the **Numeric** style:

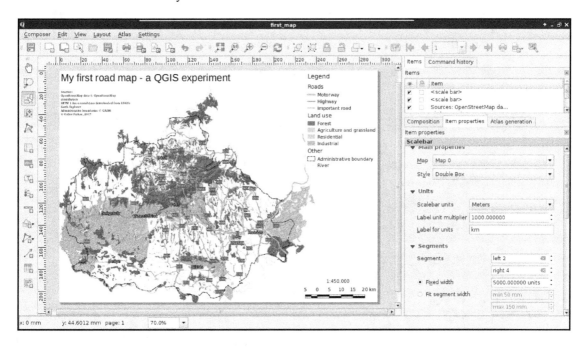

> The scale can also be added with a text box, however, adding it with a scale bar item is more convenient, since if we change the scale of our map, we don't have to update it manually.

As a bonus task, let's add one final element to take up the empty space between the legend and the scale--an inset map. By adding additional data frames, we can focus on smaller areas in greater detail. Let's choose an area we would like to emphasize, and create a new data frame with the **Add new map** tool. If we resize it to fit the width of the legend and the scale bar, we end up with a large map and a small map showing exactly the same area. However, by using the **Move item content** tool, we can zoom and pan our new map to fit our needs.

> If we have two or more maps, they will most likely have different scales. QGIS does not know which map we would like to use for the scale bar items, therefore, it allows us to specify it under the scale item's **Map** property.

QGIS offers a very handy tool for showing the extent of a data frame on another data frame. To access this property, let's select the large map's item, and navigate to **Overviews**. If we add an overview with the plus sign, and specify the reference to our second map in the **Map frame** property, we can see the extent of our inset map showing up on our main map. We can customize the look of this extent in the **Frame style** property:

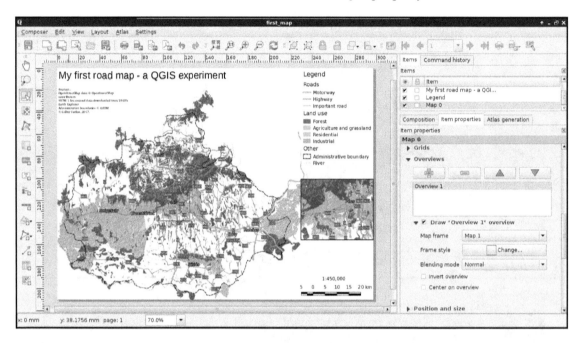

Wondering how I got the projected lines in the map from the first chapter? Well, manually of course. You can use the **Add Nodes item** tool to add additional lines and polygons directly on the canvas.

The only thing we did not add to our map is the north arrow, as our map is oriented towards North. Unfortunately, adding a north arrow is far from trivial in QGIS 2. The first step is to add an image with the **Add image** tool. Under the image item's properties, we can find the **Search directories** menu, which contains some of the default SVG images shipped with QGIS. Among them there are some north arrows. The only problem remaining if we have to add a north arrow is that our map is rotated. An image item, on the other hand, is not. To solve this problem, we can check the **Sync with map** checkbox in the **Image rotation** menu. If our map is not rotated by hand, but by the CRS used, we can use the **True north** option in **North alignment** (*Appendix 1.2*).

Summary

Congratulations! You have just created a nice map. Of course, it has some more or less obvious flaws, but I would have been very pleased if I could manage to create such a map back in my school days. Some of the flaws are more obvious, like polygons sticking out from the administrative boundary. Well of course, we can consider it artistic, but a proper map either clips its content to an irregular shape, or continues to show its thematic beyond it. We will fix that in the next chapter. Less obvious flaws are the occasional dangling lines disconnected from the visualized river and road networks, or the ill-placed labels, which would be quite hard to correct from QGIS. However, don't worry about these for now; just enjoy the feeling that you have just created a great map.

If you followed the entire chapter in one sitting, take a rest. That was a lot of knowledge compressed into a single chapter. Let it sink in. In the next chapter, we will dive into the flaws of our map, and try to correct some of them with QGIS if possible. We will learn about some of the possible export formats of digital maps, and their usefulness. Then we will learn about other spatial data exchange formats by exporting our layers, or just parts of them.

5
Exporting Your Data

In the last chapter we learned how to utilize the representation model of GIS to create spatial visualizations and digital maps. Our final map is very nice, although it has some flaws remaining. In this chapter, we will fix some of the issues (for example, dangling lines, features sticking out of the study area) and learn how to export our map to two kinds of graphics--**SVG** (**Scalable Vector Graphics**) and image formats (for example, PNG). Then, we will go on to learn about different spatial data exchange formats. We will discuss their main properties and limitations to be able to choose the best for our project.

In this chapter we will cover the following topics:

- Clipping vector and raster data
- Exporting as graphics
- Spatial data exchange formats

Creating a printable map

As we saw in the previous chapter, even the simple task of making a map can involve some geoprocessing. Now we will dive further into using basic geoalgorithms to fix some of the more obvious flaws of our map before exporting it. The most basic geoalgorithms in GIS are *buffer, clip, intersection, difference, union,* and *merge.* Some experts also consider dissolve a basic geoalgorithm, although it involves merge and multipart conversion with some criteria. Now we will use clipping to get features only in our study area. Clipping is similar to extraction used in the previous chapter, although it not only selects features within a mask but also clips the features to its boundary. It's like placing a cookie-cutter shaped as our study area in our case and only keeping the parts underneath.

As we use more and more geoalgorithms, the instructions will be less informative and more concise. For the first few, we will discuss accessing and parameterizing the whole algorithm.

Clipping features

In order to clip our layers we need two things--an *input layer*, and a *clip layer*. The clip layer contains the cookie-cutter, while the input layer is clipped to the clip layer's shape. As we have a filter on our administrative boundaries layer, we have the perfect clip layer at hand. Any vector layer can act as an input layer, however we should only consider layers sticking out from our study area. For every input layer, we have to iterate the following steps one by one:

1. Select **QGIS geoalgorithms** | **Vector overlay tools** | **Clip** from the **Processing Toolbox**.
2. Select the current input layer as **Input layer**.
3. Select the filtered administrative boundaries layer as **Clip layer**.
4. Type memory: as the output name.
5. After running the algorithm, copy the styling of the original layer to the new one with the two simple steps from the previous chapter.

After running the algorithm, we get the clipped version of our input as a memory layer. This is the first format we discuss. It is a very important, very handy format for storing intermediate results. Memory layers differ from temporary layers, hence the latter are saved to a temporary folder as shapefiles and deleted once we restart our computer. Memory layers can be edited in place; we can do any modifications we see fit before saving the final results to the disk. There is only one thing we cannot do with memory layers--save them with our project. Once we close QGIS, the memory layers are gone for good, only their **Layers Panel** entries remain:

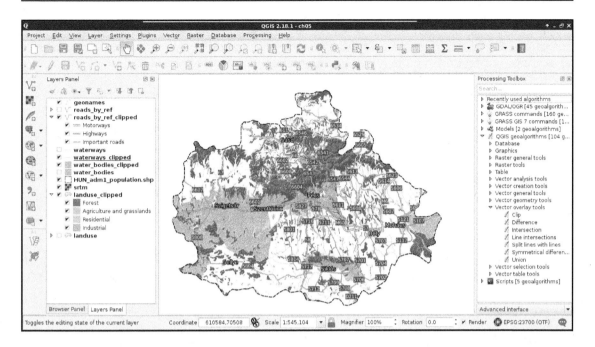

You don't have to give a name to the layer after the **memory:** notation. In the current version, memory layers inherit the algorithm's name, thus it is strongly recommended to rename the layer immediately by right clicking on it in the **Layers Panel** and selecting **Rename**. Furthermore, you can save memory layers in your project file with the `Memory Layer Saver` plugin. It saves memory layers in a binary format to an `mldata` file along with your project file.

Creating a background

The next flaw we correct is much less obvious and could be argued to be an error. The background of our data frame is exactly the same as the rest of the paper. This can introduce some ambiguity, which we can resolve by changing the background in our study area to another color and creating an additional **Other** category for land use in our legend. The easiest way to do this would be adding a fill to the administrative boundary and pulling it down to the bottom of the layer list, therefore the rendering pipeline.

The problem with this approach is our `srtm` layer gets blended into these new areas, creating a lot of noise. Fortunately, with some clever processing, we can create the negative of our land use layer using the following steps:

1. Select **QGIS geoalgorithms** | **Vector overlay tools** | **Difference** from the **Processing Toolbox**.
2. Supply the filtered administrative boundary as **Input layer**, and the land use layer as **Difference layer**. Save the result to a memory layer.
3. Pull the result just above the `srtm` layer, and style it using a very light color with no outline.
4. In the new layer's **Style** tab, select the **Fill** parent category. Next to **Color**, select the interactive color chooser with the down arrow next to the field. Click on **Copy color**.
5. Open the land use layer's **Style** tab, add a new category with the plus button.
6. Set the **Label** to **Other**, the **Filter** to **FALSE**, and paste the copied color using the interactive color chooser of its **Fill** parent style category. Finally, remove the outline from the style:

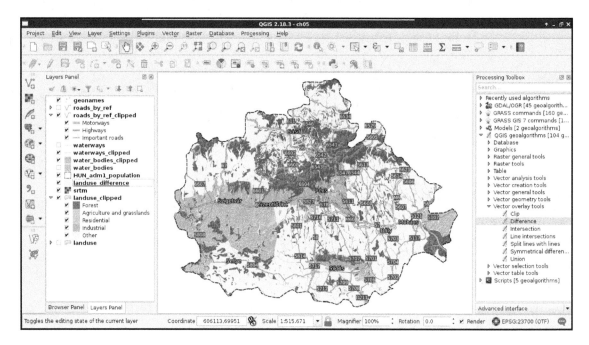

The **Difference** tool returns the difference between the geometries of the first layer and the second layer. It basically erases the second layer from the first layer and returns the results. Some of QGIS's geoalgorithms are error-prone and even can cause a crash. For example, the **Difference** tool crashed for me. If this happens, we can always choose GRASS GIS's equivalent algorithm. GRASS GIS is a professional, very stable, and quite fast software, although it has a quite steep learning curve and assumes its users are proficient GIS users with some programming knowledge.

 Don't worry about installing or configuring GRASS. We already installed it in `Chapter 1`, *Setting Up Your Environment*, amongst other GIS packages, and QGIS can automatically access some of GRASS's functionality from then on.

We will discuss the peculiarities of GRASS GIS in a later chapter. For now, the required tool for achieving the same result is **v.overlay**, which can be accessed from **GRASS (GIS 7) commands** | **Vector** in the **Processing Toolbox**. This tool is a great example of GRASS GIS's philosophy. It requires two input layers (**A** and **B**), and we must be able to distinguish between them. The **A** layer is the input layer, while the **B** layer is the reference, or mask layer. Therefore, we need to specify our administrative boundaries layer for **A** and our land use layer for **B**. This tool groups some of the mostly identical basic geoalgorithms. The specified operator decides which algorithm it should run. The AND means clip, the OR means union, the NOT means difference, while the XOR means symmetrical difference (*Appendix 1.3*). After we define **not** as the operator, we only have to choose an output type, which cannot be a memory layer for any GRASS algorithm. We can either save the result to a temporary file or specify an output. As the two approaches are almost the same (disk usage occurs in both cases), specifying the output is recommended as we can restore it later from the saved project.

Removing dangling segments

One of the perks of using memory layers is we can edit the features before saving them to disk. We can delete disconnected lines or add missing ones. This is a task which can be hardly automated. With sophisticated algorithms we could treat the road layer as a network and find disconnected parts automatically. However, it is a quite cumbersome approach in QGIS, while manual editing is feasible for such a low amount of features. Furthermore, in the rivers layer, it is far from trivial which parts belong to the main network as we would need the whole waterways layer for network analysis. If we select every feature shorter than an arbitrary threshold and delete them with the **Delete Selected** tool, we can easily remove connected but short segments.

TIP

If you would like to delete parts from the roads layer, use the **Edit** | **Delete Part** tool from the main menu. We dissolved the roads layer, therefore we have multipart geometries. If you select a part and remove it, QGIS will remove every road with the same road number.

We can also add some new features. For example, the Geofabrik extract does not contain water transports. If we have a ferry connecting important roads, we can add that feature manually and modify the road layer's style to show the new class:

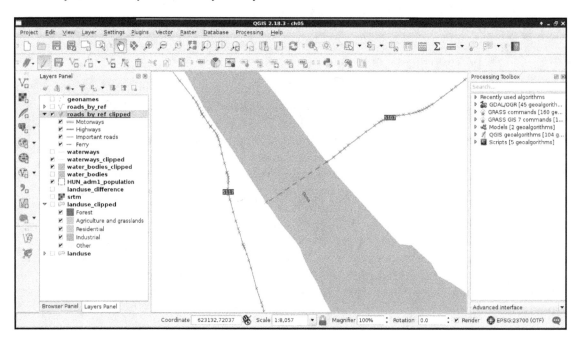

Exporting the map

Now that we have included the modifications our map needs, we can export our map from the print composer. As we have a composer in our project, we can access it with the **Composer Manager** tool from the main toolbar. If the maps do not show anything, we can select their respective items and click on **Update preview**. The only thing we have to modify is the legend. The bad news is we need QGIS to recreate the whole legend with the legend item's **Update all** button and modify the result. Fortunately, this is quite an easy task:

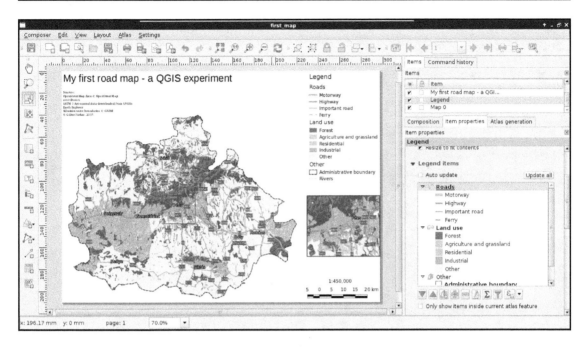

If the result looks good enough in its current form, we can export it directly to an image or PDF format. We can access the tools **Export as image** and **Export as PDF** from the main toolbar of the print composer.

A good way for post-processing - SVG

After creating the best visualization possible in QGIS, we might still need some changes to be made. This process is called post-processing and can be done in various ways. We can edit the map in an image editor, or we can export the map in SVG format and edit it with a vector editor. With Scalable Vector Graphics, we can drop georeferencing but still keep our vectors using screen coordinates. Let's do that by exporting the map with the **Export as SVG** tool. On selecting the destination of our SVG file, a dialog pops up offering some options. We should check the **Render map labels as outlines** checkbox. This way, we can move labels easily.

If you open the result in a vector editor (such as Adobe Illustrator or Inkscape), you will realize that the blending options are gone. The SVG format does not support storing blending modes and every software uses its own mechanics for this. As QGIS exports different layers to different SVG layers, we can access the `srtm` layer (we have to fiddle a little bit to find it) and apply a blending mode available in the software.

SVG export can create some artifacts (unwanted or erroneous rendering), but they can be handled by masking out the unwanted parts:

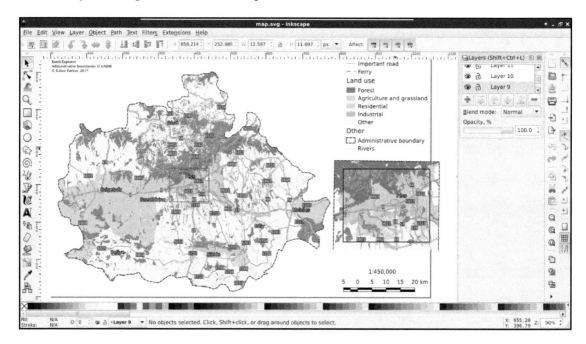

 The two main purposes for which I use the SVG format is manual label management (moving and removing labels) and converting lines with sharp joins to curves (if it is feasible with the given thematic).

Sharing raw data

Being able to create digital maps from data is a useful skill, although in most of the time, we are using spatial data exchange formats to save our raw data. We are using these formats to save our edits, export a subset of our database to use it locally, or share our data with others. There are a vast amount of different spatial formats, and all of them has their peculiarities, their most appropriate use cases. We should be able to choose the most appropriate format for our case.

Vector data exchange formats

There are a wide variety of vector formats with quite distinct purposes. Some of the vector formats are created to fully support the specialities of the host software while others are more general ones. A unique property of vector GIS formats is that there isn't an all-purpose, all-in-one format. There are two main types of formats--binary and ASCII (text-based). Binary formats are more concise; however, they require parsing algorithms with binary magic directly developed to support them.

ASCII formats, on the other hand, are human-readable and some of them (like JSON and XML) are natively supported by several high-level programming languages (especially, fourth-generation languages). Others can be interpreted by easily implementable string manipulation techniques.

We can export layers from QGIS by right-clicking on their items in the **Layers Panel** and selecting **Save As**. In the dialog, we can not only specify the new file's destination, name, and format, but we can also save only selected features (if there are any) by checking the **Save only selected features** checkbox. If we would like to exclude one or more of the attribute columns, we can expand the menu **Select fields to export and their export options** and specify the columns that we would like to export. Finally, we can also specify a very important property of the vector layer--its CRS. If we select a CRS different from our layer's, QGIS automatically transforms our features to the selected destination CRS.

Shapefile

Shapefile is one of the oldest vector formats around. It is a general binary format supporting simple features (points, lines, polygons, and their multipart counterparts) and their attributes. It is widely supported by GIS software. The format was initially created for ArcView 2 and later open sourced by Esri. If we need a format which is most likely supported by the destination software, Shapefile is a safe choice. Shapefiles can only store one layer and every layer stored as a Shapefile usually consists of four different files. The shp file (which we browse if we manually open a Shapefile in QGIS) stores the geometry data in a binary format, the prj file stores the projection of the layer, the shx file stores indexing information for faster lookups, while the dbf file stores the attribute table of the layer in a relational style.

There is a similar, not much less popular semi-proprietary format--the MapInfo TAB created for the GIS software MapInfo Professional. If you come across such a layer, you can open it with QGIS by browsing in the **tab** file.

Since Shapefile is a 25-year-old format, it has more limitations than advantages. It cannot store topology, and neither the geometry file nor the attribute file can exceed 2 GB of physical size. These are the kinds of limitation we could live with, as the format is fast and widely supported. However, the biggest limitations come from the database format it uses in the dbf file--dBASE. There are numerous limitations of the old dBASE format, including a maximum field name (column) length of 10 characters, only three supported attribute types (integer, floating point, and text), and supporting only 255 fields per layer.

You can take a look at a dbf file with a spreadsheet editor supporting old dBASE formats, like LibreOffice Calc.

As the Shapefile format has numerous limitations, discussion about a competent alternative is a trending topic among GIS users and developers. Currently, there are some competitors taking the historical place of shapefiles, although there isn't a single all-purpose widely supported format yet. They are all capable of storing multiple layers in a single structure and do not have practical size limitations (140 TB for GeoPackage), neither in geometries, or in attributes. The portable SQLite database can store multiple vector layers in a single file, which can be accessed by SQLite and its spatial extension--SpatiaLite. It does not support topology, raster storage is not straightforward, and it is more of a general-purpose self-contained DBMS than a spatial format. The File Geodatabase by Esri is well-tailored to the ArcGIS data structure and can hold vectors, rasters, other type of layers (like topology layer) and even scripts in a single structure. However, it is not an open standard and has a limited, read-only compatibility with GDAL and, therefore, with QGIS. There is also a new standard with vector and raster support built on top of the SQLite format--the GeoPackage. It might become the general open source format of the future, although it still misses topology support and as it is a 2016 standard, it is not widely supported by GIS software. GDAL can read and write it, so you can try it out in QGIS.

WKT and WKB

WKT is short for Well-Known Text, while WKB is the abbreviation of Well-Known Binary. They are the ASCII and binary representations of the same format, mainly used by spatial extensions of (O)RDBMS software (such as PostGIS). They support 2D geometries and introduce some special types besides the simple ones (for example, curve, TIN, surface, and others). We can come across WKT representations of geometries mainly in official PostGIS and QGIS examples, as both like to use it. A WKT representation of a point looks like the following:

```
POINT (17.80554 46.04865)
```

The format is rarely used with files (it cannot even represent attributes), however, it is a very common format to communicate geometries to SQL-based software. There are extended versions of these formats called EWKT and EWKB, which were created by the team behind PostGIS, and are mainly used in it. It extends WKT and WKB by standardizing geometries in higher dimensions (up to four).

Vector geometries in GIS usually can take up four dimensions. They are called X, Y, Z, and M. The first three are unambiguous, while the fourth is usually called the measure coordinate. For example, in a 3D river streamline, the first three coordinates represent the vertices in 3D space, while the M coordinate can be used to store river kilometers (or river miles), the distance from the mouth along the river.

Markup languages

There are three notable spatial markup languages which use the **XML (Extensible Markup Language)** specification--**GML (Geographic Markup Language)**, **KML (Keyhole Markup Language)**, and **OSM XML**. All of them are XML-based formats, therefore, they are very verbose but also well-structured. Let's export our filtered GeoNames layer to GML and KML as discussed in the beginning of the chapter.

In the KML format, attributes are not exported automatically. You have to use the **Select All** button in the **Select fields to export and their export options** menu to include them.

If we open the exported files with a text or code editor, we can see that they are structured according to the XML specification. There are tags enclosing geometries, properties, and other information. The main difference between the two files is the type handling. In the KML file, the type of every column is defined near the end of the file, while we cannot see any type definitions in the GML file. As the KML specification was popularized by Google to offer a data exchange format for Google Maps and Google Earth, those files are self-containing.

That is, all information is stored in a single file, which can be parsed by the desktop application Google Earth and the web application Google Maps in the same way:

GML, on the other hand, was created to support web applications. Therefore, the intended (and most common) usage is via a server application (like MapServer or GeoServer) providing an XSD schema additionally linked in the GML response. The XSD schema contains metadata, like the bounding box of the layer, its projection, and the types of the columns. As we are now working in a sole desktop environment, we don't have such a server creating a schema for the exported layer. To overcome this issue, GDAL creates a `gfs` file containing this information.

The last one is a very unique format in multiple aspects. First of all, QGIS cannot write into OSM XML as it is used to export and hold a smaller amount of data from the OpenStreetMap database. We can get data in such a format directly from OpenStreetMap by zooming in on a sufficiently small area and using its **Export** tool. On the other hand, QGIS can read OSM XML, although the workflow is far from trivial. If we have an `osm` file, we can use the **Vector** | **OpenStreetMap** | **Import Topology from XML** tool to build an SQLite container from the OSM data. Next, we have to use the **Vector** | **OpenStreetMap** | **Export Topology to SpatiaLite** tool to build accessible layers from the OSM data in the SQLite container. We can only access the layers after the second step if we exported the data correctly.

We have to walk through this cumbersome workflow every time we process an OSM XML file due to its other peculiarity--OSM XML is the only markup language that stores topology. It reflects and inherits the OSM data model; it is fully topological. In the OSM database, every point is a node, lines and polygons (ways) consist of nodes, while other objects (relations) consist of ways and nodes (for example, multipart geometries). As a result, every vertex is only stored once, while every other occurrence refers to it. As OSM XML inherits this topological vector model, we have to build the topology in the first step in an SQLite container, while we also have to build layers with geometries accessible by QGIS in the second step:

GeoJSON

The last format that we'll now discuss is the GIS data exchange standard built upon the famous JavaScript exchange format--**JSON (JavaScript Object Notation)**. GeoJSON is a very permissive format inheriting the object-oriented nature of JavaScript. Type and shape consistency is not required in the format; we can have as many attribute types in a column and as many geometry types in a layer as we want. By definition, the permitted types are integer, floating-point number, text, boolean, null, array, and object. The permitted geometry types are point, line, polygon, and their multipart versions. Let's export our filtered GeoNames layer to GeoJSON.

As we can see, the result is much more concise than the markup languages but we can still interpret the features as GeoJSON is also an ASCII format. It is mainly used by web applications to visualize static vector data. It is favored for its small size and, therefore, the smaller traffic it generates on being read by the client application.

 There is a topological variant called TopoJSON. TopoJSON not only stores the vector data topologically but also quantizes the coordinate values (stores them as integers), creating smaller files. Currently, GDAL is able to read TopoJSON files with the GeoJSON parser but it cannot write to this format.

Raster data exchange formats

Similar to vector data exchange formats, there are a lot of raster formats out there. There are formats storing rasters in binary form and in ASCII form. We can even save rasters like vectors in QGIS with the **Save As** tool accessed from the raster layer's context menu in the **Layers Panel**. If we see the save dialog with the **srtm** layer, it is slightly different from the vector version reflecting the specialities of the raster data model. We can set a CRS and ask QGIS to transform the raster to another CRS (it is called raster warping). We can also set the extent and the resolutions in both dimensions manually. There are some other options too; however, they are GeoTIFF-specific. What is very unusual, though, is that we can hardly access any format other than GeoTIFF from this menu.

GeoTIFF

Unlike vector data exchange formats, we have a general, all-purpose, widely supported raster format--GeoTIFF. Due to the existence and popularity of the highly capable TIFF specification for storing lossless image data, it was extended to store spatial data along their metadata (for example, projection and georeferencing). There are only two drawbacks of this format--its size and the web. As it only supports a few compressing methods, its lossy JPEG compression cannot race, for example, with the JPEG2000 standard if data loss is acceptable. Furthermore, web browsers cannot handle TIFF files; therefore, rasters have to be converted to regular images before using them in a web client application.

GeoTIFF offers a lot of options to work with. It can create internal pyramids for various resolutions for faster visualization. It can also utilize internal tiling to speed up processing when the raster is very large. It can even compress data with various algorithms.

There are some lossless options like LZW or deflate (ZIP), and there is the lossy JPEG compression for a significantly smaller file size.

> When you save a raster image with QGIS's **Save As** dialog, you can specify the compression in the **Create Options** menu's **Profile** field. Choosing **High compression value** results in a GeoTIFF image compressed by the deflate algorithm.

Clipping rasters

Until now, we only used vector processing, so let's see how raster processing is different in QGIS. Let's clip the SRTM raster to our study area. We can do it by following these easy steps:

1. Select **Raster | Extraction | Clipper** from the menu bar.
2. Specify our `srtm` layer as the input layer.
3. Specify the output directory and the file name, and select `GeoTIFF` as the format.
4. Check the **Mask layer** checkbox for the **Clipping mode**.
5. Select our filtered administrative boundary layer as the **Mask layer**.
6. Run the algorithm.

Did the algorithm finish successfully? If it did, you can see the clipped layer on the map canvas. If not, don't worry, it didn't work for me either. Before fixing the issue, let's see how QGIS uses the raster processing tools. Unlike vector processing tools, QGIS has only a few of its own raster tools. It uses external tools to achieve raster processing, usually, via command line or Python. The tool we opened is a clever wrapper around GDAL's warp tool, whose main purpose is warping rasters between CRSs. But it can also clip the input raster in the process. We can even see the command that QGIS uses at the bottom of the window:

```
gdalwarp -q -cutline "/home/debian/practical_gis/processed
/admin_boundaries/HUN_adm1_population.shp|layerid=0|subset="NAME_1" =
'Baranya'" -tr 0.000277777777778 0.000277777777778 -of GTiff
/home/debian/practical_gis/processed/srtm/srtm.vrt /home/debian
/practical_gis/processed/srtm/srtm_clipped.tif
```

This architecture implies a very important specificity of raster processing in QGIS. We must have a physical copy of the data we use. In raster processing, memory layers are obsolete-- we cannot even select them as mask layers in the **Clipper** tool.

Raster (and image) processing was always a somewhat special niche in GIS. Specialized software used to have more capabilities than general GIS software both in the commercial and open source worlds. Until recent times, even Esri couldn't monopolize image processing like it did with GIS. ERDAS Imagine, ENVI, and eCognition, even today, can give more complete solutions for raster-based workflows. This is the same as in the open source segment; there are specialized tools for these tasks. QGIS, on the other hand, does not have to reinvent the wheel. It can utilize some of the greatest open source tools for processing raster data (Orfeo Toolbox, GDAL, GRASS GIS, and SAGA) if installed and configured properly.

The problem with our approach was the filter on our administrative boundary layer. QGIS couldn't manage to include the filter properly in the command-line call it created. To solve the issue, the easiest way is to export our administrative boundary layer. As we applied a filter, only one polygon gets exported. After the export, we can specify the new layer as mask and run the algorithm again:

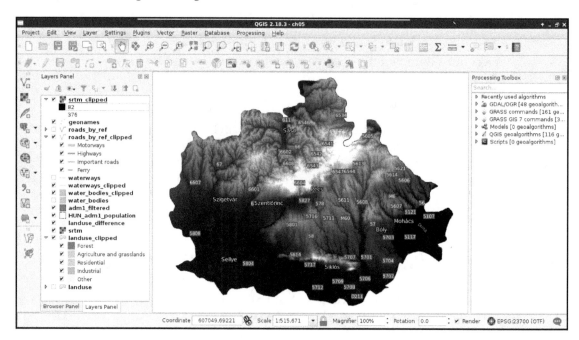

The other, slightly more advanced solution is to consult the documentation of GDAL's warp tool at `http://www.gdal.org/gdalwarp.html` and modify the call manually with the pencil-shaped **Edit** button. In the documentation, you will see that a filter on the cutting shape can be applied with the `-cwhere` parameter and a basic, regular SQL expression. Therefore, the same call should look like the following:

```
gdalwarp -q -cutline "/home/debian/practical_gis/processed
/admin_boundaries/HUN_adm1_population.shp" -cwhere "NAME_1 =
'Baranya'" -tr 0.000277777777778 0.000277777777778 -of GTiff
/home/debian/practical_gis/processed/srtm/srtm.vrt /home/debian
/practical_gis/processed/srtm/srtm_clipped.tif
```

If you right-click on the clipped `srtm` layer and select **Properties** | **Transparency**, you can uncheck the **No data value** box. Then you can see that the resulting raster inherits the size of the source raster and fills the excess cells with zeroes. To crop the raster to the cutting shape, you can check the **Crop the extent of the target dataset to the extent of the cutline** box in the **Clipper** tool.

Other raster formats

We could see when we chose our destination folder that there are a lot more raster formats which QGIS can export to via GDAL and other tools. Most of the time, GeoTIFF is sufficient as it can use various compression methods, handle big rasters with the internal BigTIFF format, and use various data types. There are some cases, on the other hand, when GeoTIFF is not an appropriate data exchange format. For example, when I tried to export raster data from IDRISI 7 in GeoTIFF, QGIS could not handle the output. There might have been a problem with IDRISI's GeoTIFF implementation or a bug in QGIS's (GDAL's) GeoTIFF parser--it didn't really matter. GeoTIFF wasn't the appropriate data exchange format for that case and I had to find another one.

From the vast number of raster formats, some of the more widely supported ones are ERDAS's **img** format and Esri's various binary and ASCII data grid formats. These formats can include some auxiliary files besides the main data file for metadata. For example, the SRTM data we downloaded are in **bil** files, which is a binary raster format mainly used to store satellite imagery. It is categorized under the **ESRI .hdr Labelled** group in QGIS, as it comes with an ASCII header file containing some general information about the raster (for example, it uses the BIL format).

Summary

In this chapter, we learned about the most popular spatial data exchange formats for both vector and raster data. We learned not only their specialities but also how QGIS can create them. We also know now how to export our digital maps both as a regular image and in SVG for post-processing. Finally, we learned how to shrink our data or enhance our map by reducing the layers to only cover the relevant parts.

In the next chapter, we will look into the definition and types of databases and how spatial databases compare to them. We will discuss the possibilities of building our spatial data structure and build our very own database with PostGIS and QGIS. Additionally, we will see how QGIS communicates with PostGIS and how we can build basic PostGIS queries.

6
Feeding a PostGIS Database

In the previous chapter, we learned about data exchange formats, which can be used to store our layers on our hard drive persistently in some files. Data exchange formats have the clear advantage of portability; however, we end up with files representing layers scattered through various folders. This isn't a problem when we have only a few layers or a well-designed folder structure, but it is not feasible for larger amounts of data. There is a solution for storing larger amounts of structured data in one easily accessible place--build a database. In this chapter, we will learn about databases in general, spatial databases, and how we can create a PostGIS database from QGIS easily.

We will cover the following topics in this chapter:

- Spatial and non-spatial databases
- Creating PostGIS tables from QGIS
- Spatial indexing
- Visualizing PostGIS tables in QGIS

A brief overview of databases

Let's start with a very brief introduction of databases. You might have more than enough knowledge about databases; if that is the case, you can safely skip to spatial databases. First of all, what is a database? A database is a collection of structured or semi-structured data, which can be, at least, updated and queried by the Database Management System (DBMS) or the library using it. Besides its very trivial benefit of storing a lot of data in the same place, the wrapper system usually offers methods for not only retrieving but also aggregating, filtering, or joining data. Furthermore, most of the DBMS and libraries are very well-optimized for their use cases, and therefore, offer faster solutions than working with traditional files and system calls.

Relational databases

The first and oldest database types are relational databases. They hold data in a well-structured form in tables. Tables consist of rows and columns. A column represents a single attribute of the data, which is stored as a specific data type. Rows represent a single data record, such as a customer's contact details, or a vector feature. Relational databases are very often transactional databases (I could only name MySQL versions released before 2010 which are not transactional). The philosophy behind transactional databases can be expressed with the anagram **ACID** (**Atomicity, Consistency, Isolation, Durability**). These are the four very important properties that relational databases offer. These properties can be explained as follows:

- **Atomicity**: There are only full transactions, no partial ones. If one part of the transaction fails, the whole transaction fails.
- **Consistency**: A transaction can only occur if it satisfies every constraint of the database. A field update for a unique, not-null, integer column must satisfy those three constraints in order to succeed.
- **Isolation**: Concurrent transactions are executed as if they were sequential transactions.
- **Durability**: Transactions are only successful if they are written to the disk. If something prevents a part of the transaction from saving it permanently, the whole transaction fails and the database is rolled back to the prior state.

As relational databases are basically a collections of tables (concerning only the stored data), this model highly resembles the vector data model. That is, vector data can be stored in relational databases quite painlessly, while storing and using raster data efficiently is more complicated. Although following the ACID principles makes relational databases very reliable for using both on personal computers and servers, their architecture makes them hardly scalable (for example, they are not the best choice for big data or very complex analysis). Furthermore, creating a good relational database requires some designing and considerations.

Relational databases are especially vulnerable to redundancy. Redundancy is not always a bad thing, for example, servers rely heavily on it to store data safely and to restore it entirely in case of a disk failure. However, redundancy can be harmful in a database. If we store the same columns in multiple tables, we can speed up queries, although we use up a lot more physical space and also make the database more vulnerable to corruption (for example, if we modify only one or a few occurrences of a redundant value). However, eliminating redundancy is not the best idea for every scenario. We can end up with a lot of tables storing only little chunks of information that we have to manage. It is a sensitive and, in most cases, subjective task to find the best normalization level (normal form) for the given relational database.

NoSQL databases

The vague termed **NoSQL** (**not only SQL**) database groups object-oriented databases. There is a very wide palette of different architectures (like document stores, column stores, graph databases, and so on). However, they have at least one common point--they store data as structured but not necessarily consistent objects. They are often not ACID-compliant, although they also offer reliability and stability in their own ways. As they store objects (not tables), stored data don't have to be consistent while redundancy is not a great issue.

NoSQL databases are generally used for different purposes than relational databases; therefore, a direct comparison is impractical. In general, object-oriented databases offer better scalability so they are almost exclusively chosen for big data over relational databases. They often use query languages other than SQL but it is also common to use APIs and function calls for querying and processing data. Finally, a lot of NoSQL DBMSs are vertically extendable, which means we can cluster them effortlessly and extend the servers they can use by putting more machines into the rack, giving them more resources.

Based on the previous statements, NoSQL databases are superior to relational databases in almost every aspect. So why should we even consider using relational databases? Well, NoSQL databases have several drawbacks--they are very difficult to set up, configure, and fill with data. Furthermore, it is harder to enforce data integrity, and the lack of transactional support in some NoSQL DBMSs make concurrent editing harder. With smaller datasets (that is, a few GBs), NoSQL databases give only a negligible performance boost. Thus, if relational databases can give a good fit for our data, setting up a NoSQL database is just not worth the time and effort.

Spatial databases

What makes a database spatial? The answer shouldn't be searched for in the database but in the host DBMS. Spatial database management systems have some kind of spatial vision. They recognize spatial data and can offer some kind of dedicated functionality involving them. They have some exclusive validation methods for spatial data. Spatial functionality can be implemented at so many levels. Some of the spatial modules in DBMSs can only recognize points, lines, and polygons, while others offer full-scale GIS functionality even for raster data.

From the relational family, the two de facto standards in the open source segment are PostGIS and SpatiaLite. SpatiaLite is built upon SQLite, which is container-based, offering portable, single-file databases. It does not offer a full DBMS with some kind of interface; it is API-based. That is, it offers a lightweight library written in multiple languages to communicate with SQLite containers. It is a lightweight but robust way to have relational databases in any software. In the end, it is left to the software to use this library or not.

The other relational DBMS is PostGIS built on top of the PostgreSQL DBMS. PostgreSQL is an **ORDBMS (Object Relational Database Management System)**, which means it can output objects instead of the traditional tabular format. This property makes it very convenient to create extensions for it. One of the many extensions that PostgreSQL has is PostGIS, which is not only a spatial index and some vector geometry types built on top of a DBMS, but a full-scale headless GIS. PostGIS is the most complete spatial database system up to date and as it is directly on top of the underlying database, its performance is a great argument in its favor.

If we have additional criteria, like strongly inconsistent data or big data, NoSQL spatial databases can come to the rescue. A lot of NoSQL DBMSs have spatial capabilities, although in most cases, they are only minimal vector-type support and spatial indexing. These specify the main spatial functionalities of NoSQL databases--using spatial data types and offering spatial queries. Some examples of spatially-enabled NoSQL DBMSs are the following:

- **Document stores**: MongoDB and CouchDB (GeoCouch). Fast, but not faster than PostGIS. We can use them for smaller inconsistent data or when we already use those DBMSs.

- **Graph databases**: Neo4j and OrientDB (partially graph). Fast, scalable, but with limited functionality. OrientDB has the closest syntax to PostGIS as it also implements the ISO/IEC 13249-3 standard.
- **Column stores**: Geomesa and GeoWave. Highly scalable DBMSs for storing and analyzing spatiotemporal big data. Geomesa is generally faster, especially in case of complex analysis, while GeoWave takes up less disk space.
- **Array database**: Rasdaman. A flexible and scalable DBMS built for using array-based (raster) big data. It supports dimensions beyond the spatiotemporal data frame.

> NoSQL DBMSs are just getting their heads around spatial capabilities. As the demand for analyzing spatial and spatiotemporal big data increases, there will be more and more complete NoSQL GIS solutions. Keep your eyes open for these systems.

Importing layers into PostGIS

PostGIS has a very big benefit over other DBMSs--QGIS can communicate with it very well. The same applies for SQLite containers and the GeoPackage format, but PostGIS offers the best capabilities as it can be deployed as a server and accessed from multiple clients even concurrently. For accessing a PostGIS database from QGIS, first we need to connect to it. Remember the database we created in Chapter 1, *Setting Up Your Environment*? We will use the data provided there to define and save a connection in QGIS:

1. Click on the **Add PostGIS Layers** button in the layers toolbar.
2. Define a new connection by clicking on **New.**

3. Fill out the required parameters we used to set up our database:

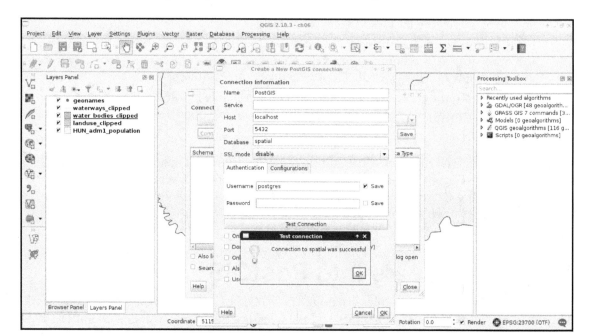

4. By clicking on **Test Connection**, we can ask QGIS if the provided parameters are correct and sufficient. If it says so, we can save the connection by applying with the **OK** button.

If you are using Windows, don't forget to provide the password you defined during installation.

Now that we have defined a connection, we can use QGIS's database manager. We can access it from **Database** | **DB Manager** | **DB Manager** in the menu bar. There we can click on **PostGIS** and select the name of the connection we defined. We can see that we have a simple PostgreSQL database as we did not enable the PostGIS spatial extension yet.

We have some permissions, though, which is nice as we can enable PostGIS directly from QGIS. To do this, we have to click on the **SQL window** button and type the following expression in the dialog:

```
CREATE EXTENSION postgis;
```

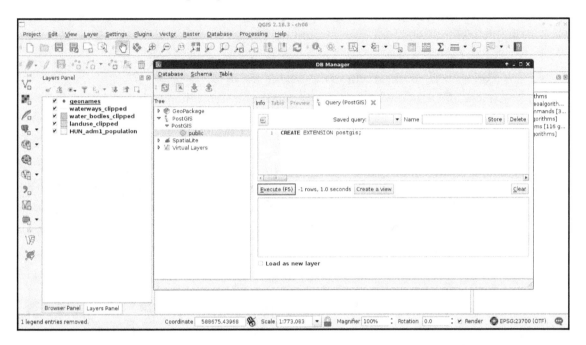

After executing and refreshing the connection (right-click on the connection's name and click on **Re-connect**), we can see some views and table created by the PostGIS extension. If we open the table `spatial_ref_sys` in QGIS and browse its content, we can take a glance at the CRSs PostGIS support (*Appendix 1.4*). Now our database is spatially enabled and we can start importing layers into it.

You can add a custom projections to PostGIS by inserting it into `spatial_ref_sys` with a **regular** `INSERT` statement. You must provide a `PROJ.4` definition of the projection, though, as both QGIS and PostGIS use the `PROJ.4` library to handle projections. The **srtext** field might be left blank (empty string), although some software might require it instead of the `PROJ.4` definition.

Importing vector data

First of all, let's load the two layers saved on the disk--the GeoNames and the administrative boundaries layers. We can import layers with the **Import layer/file** button accessed from QGIS's database manager. QGIS automatically converts the specified layer to a PostGIS table by providing the required properties. We can import virtually any vector layer with the following steps:

1. Choose the input layer in the **Input** field. We can also specify layers not loaded in QGIS with the browse button (**...**) next to the field.
2. Choose a **Schema** to load the layer into. In our case, the **public** schema is the only one available.
3. Provide a table (PostGIS layer) name in the **Table** field.
4. Define the source and target projections. It can be omitted if you would like to keep the current projection. Let's save the layers in the local projection you're using by using `4326` as **Source SRID** and your local projection's EPSG code as the **Target SRID** (without **EPSG**).
5. Check the **Create spatial index** box. Consider the following screenshot:

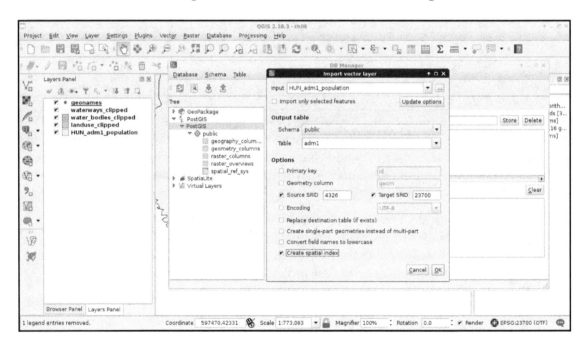

TIP

You can also import layers into PostGIS with the ogr2ogr command-line tool. If you use Windows, you can access this tool from the OSGeo4W Shell. In our case, the command should look like the following (variables prefixed with the word your): `ogr2ogr -f "PostgreSQL"` `PG:"host=localhost user=postgres dbname=spatial` `password=yourpassword" yourfile -nln yourtablename.` Password should only be included in password-protected databases or Windows environments.

If we reconnect to the database, we can see the two imported tables. We can inspect them by clicking on them and navigating to the **Table** and **Preview** tabs. If we inspect the values stored in the tables, we can conclude that every value is loaded successfully. We have null values, integers, floating point values, and strings. The only problem is that we cannot define which columns we would like to import. If we delete the columns within QGIS, we would lose them also on the disk. Luckily, we can run the SQL queries within QGIS's database manager and drop the excessive columns manually:

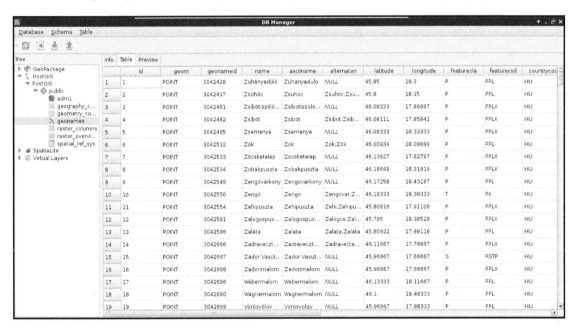

There are only two mandatory columns--id and geom. The id column stores unique identifiers in the integer format. It can be removed, although keeping it is highly recommended for easier data management. The geom column stores geometries in PostGIS's native format; if we remove it, we end up with no geometries at all. Both of their names can be altered from QGIS's import dialog.

6. By opening the **SQL window**, we can build an SQL expression for dropping columns. It does not matter what layer we selected as we must provide the table to remove columns from. The syntax for dropping columns looks like the following:

```
ALTER TABLE tablename DROP COLUMN column1,
  DROP COLUMN column2, etc.;
```

Therefore, to drop the id_1, name_0 and iso columns of the adm1 layer, we can run the following expression:

```
ALTER TABLE adm1 DROP COLUMN id_1, DROP COLUMN name_0,
  DROP COLUMN iso;
```

 As we now use PostgreSQL's SQL syntax, we don't have to enclose the column names within double quotation marks.

We had an easy task importing two basic layers into PostGIS--one with point geometries and one with polygon geometries. What happens when we have layers with mixed geometry types? Of course, mixing points, lines, and polygons are strongly discouraged, but mixing single and multipart geometries of the same type is a common thing. The only problem is that PostGIS doesn't allow us to do so.

7. Let's try to import our in-memory waterways layer:

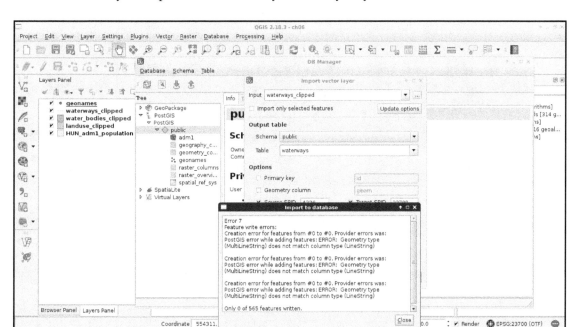

No matter how we try to import, if we have mixed geometry types, we end up with an error. Our layer has line string and multipart line string geometries mixed, while PostGIS only accepts one of them. The solution is simple though. We either have to explode multipart geometries or convert single geometries to multipart ones based on an attribute. Although we would get a little more optimal result if we stick with the lowest common denominator (in this case, multipart geometries), QGIS does a poor job in doing it. To convert our features to multipart geometries, we would need a column with duplicated values, therefore, we would lose data. As this is not an ideal solution, we have to convert our layer to single-part geometries, duplicating some of the data.

There is a tool called **Convert geometry type** in **QGIS geoalgorithms | Vector geometry tools**. It is theoretically able to convert single-part linestring geometries to multiparts, although if it finds a feature which only consists of one line, it leaves that feature as a single part.

8. We can use **QGIS geoalgorithms** | **Vector geometry tools** | **Multipart to singleparts** from the **Processing toolbox** to achieve this. The output should be a memory layer (**memory:**):

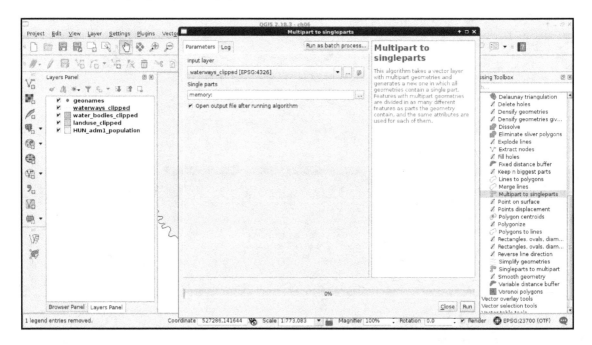

Now we can upload the resulting layer to PostGIS with the previous steps. If we put these steps and considerations together, we end up with this workflow to import every layer we need into PostGIS:

1. Clip the vector layer to the study area if it is not clipped already. Of course, you can import the whole layer but doing so will have a performance hit.

2. Try to import the layer to PostGIS. If you have mixed geometries, the import will fail.

3. Convert the multipart geometries to single part with the **Multipart to singleparts** tool. You can achieve the same with other tools; don't be afraid to experiment.

4. Import the new layer to PostGIS. Don't forget to check the **Replace destination table (if exist)** checkbox as if you have a failed attempt to create a table, an empty table will remain there. The problem is it may have the wrong geometry type.

There is still one way to convert single-part geometries in a mixed layer to multiparts while importing into PostGIS. The tool **GDAL/OGR | Miscellaneous | Import Vector into PostGIS database (available connections)** allows you to define explicit type casts by filling in the **Output geometry type** field and keeping **Promote to Multipart** checked. You still need to have the layer saved on the disk as it uses the command-line tool ogr2ogr to achieve the import. Don't forget to select your local projection both in the **Assign an output CRS** and **Reproject to this CRS on output** fields if you do so.

We will need the following layers added to the PostGIS database:

- The whole administrative boundaries layer
- The clipped GeoNames, water ways, water bodies, and land use layers
- The **OSM (OpenStreetMap)** layers we did not bother with or process before: the road layer (gis.osm_roads_free_1.shp), the railways layer (gis.osm_railways_free_1.shp), the buildings layer (gis.osm_buildings_a_free_1.shp), the POIs layer (gis.osm_pois_free_1.shp), and the transport layer (gis.osm_transport_free_1.shp)

Now that we have a bunch of layers we did not clip or alter too much to import, we need an effective way to process them. First of all, we need to decide if we would like to add them to the canvas before processing. If we do so, we can spare the overhead time of reading in the layers and furthermore, build spatial indexes on them to speed up the clipping process. We can also leave them on disk and spare some clicks. Either way, let's apply the usual filter on our administrative boundaries layer and open the **Clip** tool from **QGIS geoalgorithms | Vector overlay tools**.

There are some layers among the listed ones which are very dense. If QGIS has a hard time rendering some of them (like roads, buildings, and so on), you can uncheck the **Render** box in the status bar. You can also build spatial indexes one by one with **Properties | General | Create spatial index** to speed up processing.

In the clipping tool, we can access batch processing by clicking on the **Run as batch process** button. There we can create as many rows as we need for processing each layer with minimal hassle. Now we face another dilemma. We can save every result as a memory layer, although we cannot name them. Therefore, we need to distinguish between the results manually.

Fortunately, it isn't that hard in our case as we have one polygon layer, two line layers with very different densities, and two point layers:

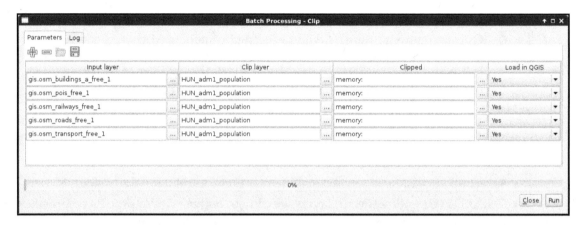

To distinguish between the two point layers, we can toggle the visibility of the original layers and the clipped layers and compare them visually. After guessing the names of the layers correctly, we can proceed and load them into PostGIS following the previous steps:

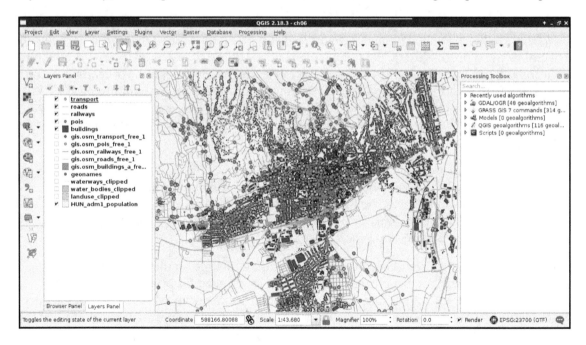

Always check the inserted layers in PostGIS with the database manager's **Preview** tab. QGIS can mix up memory layers, especially if they were saved on disk with the `Memory Layer Saver` plugin. If the inserted layer is another one, you have to reload the original one from the **Browser Panel**, clip it, and then insert it again. You can drop (remove) a PostGIS table with QGIS's database manager by right-clicking on its entry and selecting **Delete**.

Spatial indexing

We used spatial indexing several times but we have still not discussed what is it. If you know about programming, you already have an idea about indexing. If not, it is like creating shortcuts to referred items. For example, we can index a column by creating an associative array (object if you like) and storing distinct values as keys and features having them as values. This way, we reduce the time complexity of a single selection from $O(n)$ to $O(1)$.

Time complexity is describing the running time of a simple algorithm in a function of input size. The big-O notation (O) is the asymptotic upper bound by definition, although we drop the constants in the final formula as they become less and less significant while we approach infinity. In the end, we have distinct big-O notations, like $O(1)$ for constant time, $O(n)$ for linear time, $O(log\ n)$ for logarithmic time, $O(n^2)$ for quadratic time, and so on.

Spatial indexing is based on the same principle--we index geometries to reduce the time complexity of spatial algorithms. However, indexing geometries, even represented in a binary format, is unfeasible in almost every possible way. Hence, we usually use bounding boxes to index geometries. These structures are collectively called **Bounding Volume Hierarchies** (**BVH**), among which the R-tree is the most popular one. Both QGIS and PostGIS use R-Trees to index geometries. The R-tree stores the axis-oriented minimum bounding rectangles of geometries. That is, every rectangle is the smallest **standing** rectangle encompassing the whole geometry.

As the name suggests, these bounding rectangles are put into hierarchy. Bigger bounding rectangles of these rectangles get calculated based on the implementation. Simple implementations do less partitioning, balancing, and optimizations in the hierarchy, resulting in a performance hit when there are a lot of changes in the tree.

However, it does not matter how carefully the R-tree is constructed; searching it remains the same: we start by querying the biggest rectangles, iterating through smaller rectangles in them, finally arriving to the bounding rectangles of our features of interest, effectively ending up with only relevant features in sublinear time (in average case). RBush, Vladimir Agafonkin's R-Tree implementation has a great demo showing how these structures work (for 50,000 features), as shown in the following image:

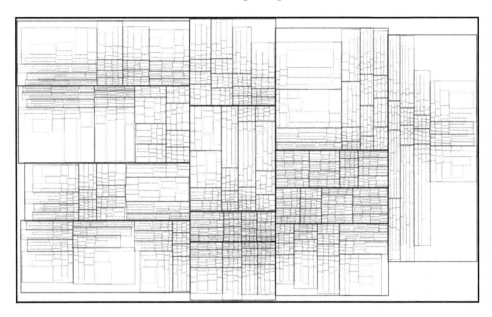

If we look at the PostGIS tables from QGIS, we can see some info about the spatial index QGIS created automatically; however, there is no detailed information. However, if we inspect one of our spatial tables in pgAdmin, we can see the exact SQL query creating the spatial index, which is as follows:

```
CREATE INDEX sidx_adm1_geom
ON adm1
USING gist
(geom);
```

So, a named index (`sidx_adm1_geom`) was created for our layer (`adm1`) based on the column containing geometries (`geom`). There is only one thing that is unclear--what is gist? **Generalized Search Tree (GiST)** is an indexing scheme which uses B-tree and, more importantly, gives a nice template to build other indexing mechanisms on. The creators of PostGIS used its extensibility to build their own R-tree implementation.

Importing raster data

While PostgreSQL's, and therefore PostGIS's, structure is optimized to use vector data, we can also store raster data in a PostGIS database. The gains in this case are not as great as with vector data as QGIS will suffer a performance hit when it has to parse raster data from a PostGIS database instead of a file. On the other hand, PostGIS can query raster tables at various locations and, therefore, we can build expressions executing spatial analysis with mixed data. Furthermore, we might need to store rasters with vectors in a distributed system. One of the main pitfalls of adding raster data to a PostGIS database is that we cannot do it from QGIS. We have to use the command-line tool `raster2pgsql`.

Our first task is to locate the tool. If you are using Linux or macOS, you just have to open up a terminal to have instant access to the tool if PostGIS is installed properly. Windows users will have a harder time, though. If you are using Windows, you have to find the installation folder of `raster2pgsql.exe`. You have to open a command line there or navigate there with `cd` commands. If we have access to the tool, we can run it without parameters and see the help page:

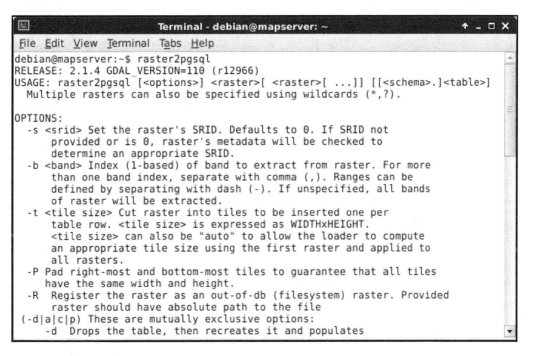

```
debian@mapserver:~$ raster2pgsql
RELEASE: 2.1.4 GDAL_VERSION=110 (r12966)
USAGE: raster2pgsql [<options>] <raster>[ <raster>[ ...]] [[<schema>.]<table>]
  Multiple rasters can also be specified using wildcards (*,?).

OPTIONS:
  -s <srid> Set the raster's SRID. Defaults to 0. If SRID not
       provided or is 0, raster's metadata will be checked to
       determine an appropriate SRID.
  -b <band> Index (1-based) of band to extract from raster. For more
       than one band index, separate with comma (,). Ranges can be
       defined by separating with dash (-). If unspecified, all bands
       of raster will be extracted.
  -t <tile size> Cut raster into tiles to be inserted one per
       table row. <tile size> is expressed as WIDTHxHEIGHT.
       <tile size> can also be "auto" to allow the loader to compute
       an appropriate tile size using the first raster and applied to
       all rasters.
  -P Pad right-most and bottom-most tiles to guarantee that all tiles
       have the same width and height.
  -R  Register the raster as an out-of-db (filesystem) raster. Provided
       raster should have absolute path to the file
(-d|a|c|p) These are mutually exclusive options:
     -d  Drops the table, then recreates it and populates
```

The next step is to prepare our raster layer. As `raster2pgsql` cannot warp rasters between projections, we have to create a transformed version if we are using a local projection. We can do this by saving the `srtm` layer in QGIS (with **Save As** from its context menu) with our local projection defined in the **CRS** field. Additionally, we can choose a **High compression** profile to decrease the file's size in the **Create Options** menu .

Now that we have `raster2pgsql` located and accessible and also our `srtm` layer exported in the projection which we would like to use in PostGIS, we can build our command for import as follows:

```
raster2pgsql -t "auto" -I -d -C /home/debian/practical_gis
    /processed/srtm/srtm_23700.tif public.srtm > srtmimport.sql
```

Let's see how the flags and parameters which we provided work:

- `-t`: PostGIS can tile raster data to speed up queries. With this parameter, we can define the size of the individual tiles. By providing the `"auto"` parameter, we ask `raster2pgsql` to calculate the optimal tile sizes for our raster. If we would like to override it with our values, we have to provide it as `WIDTHxHEIGHT` (for example, `512x512`).
- `-I`: Creates a spatial index for the bounding boxes of the tiles. It can speed up queries even more.
- `-d`: Drops the destination table if it exists before feeding our data.
- `-C`: Enforces adding constraints. Without it, the metadata of the raster (like projection) won't get added and we cannot visualize the raster in QGIS.
- `/home/debian/practical_gis/processed/srtm/srtm_23700.tif`: This is the path to our raster image.
- `public.srtm`: This is the destination schema and table name for the stored raster. We have to provide it in the form of `schema.table`.

The last part of the command redirects the output of `raster2pgsql` to a file named `srtmimport.sql`. We can use that output to create our table in PostGIS. There are additional parameters we can use. If our raster does not contain projection data, for example, we can add `-s SRID`. That is, if we would like to import a WGS84 raster, it will be `-s 4326`.

 SRS (Spatial Reference System) is a synonym of CRS. PostGIS uses the term SRS and therefore we can refer to projections used by PostGIS with their **SRIDs (Spatial Reference ID)**. You can look at the available IDs at the `spatial_ref_sys` table's `srid` column.

The last step is to execute the resulting SQL file in PostGIS. For this, we can open up pgAdmin 3 and click on the **Execute arbitrary SQL queries** button (**Tools | Query Tool** in pgAdmin 4). There we can open our SQL file with the **Open files** button. Once pgAdmin reads the content, we can run our query with the **Execute** button. That's it. If we refresh our database manager's connection in QGIS, we can see our raster layer added to our PostGIS tables:

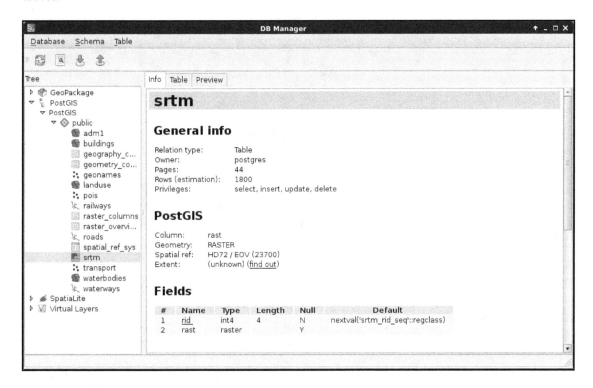

Visualizing PostGIS layers in QGIS

Great work! You just created a spatial database filled with every kind of data. Now let's visualize them in QGIS. First, get rid of the opened layers, and read the layers from PostGIS by dragging and dropping them from the database manager to the canvas. Visualizing PostGIS layers in QGIS is as easy as that.

We don't even have to worry about changes in the database. By changing the scale of the map or clicking on the **Refresh** button in the main toolbar, QGIS will automatically incorporate every change occurred since the last refresh:

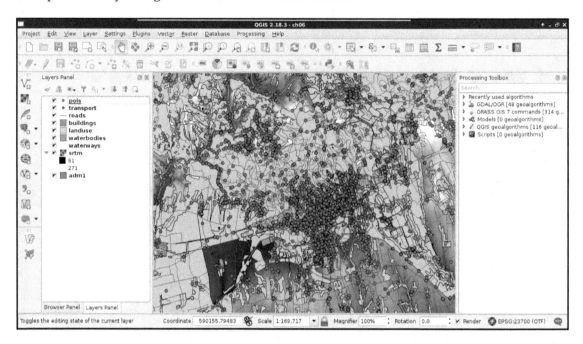

Be careful with visualizing PostGIS raster layers, especially with large tile numbers. QGIS (GDAL) will eventually make too many calls and PostgreSQL will refuse serving additional data with the error message of too many clients opened.

Basic PostGIS queries

Now that we have access to our layers in PostGIS, let's try some queries. Visualizing a whole layer from a database can involve a lot of traffic as databases are often on remote servers distributing all kinds of data. To have only the required data which we would like to work with, we can query those tables and visualize only the results in QGIS.

As a warm up, remove the administrative boundaries layer and build an expression querying it from the database.

1. Open an **SQL window** in the database manager.
2. In `Chapter 3`, *Using Vector Data Effectively*, we discussed a traditional SQL query for selecting everything from a table. Let's use the most basic select query on our administrative boundary layer--`SELECT * FROM adm1;`.
3. Check the **Load as new layer** checkbox. We can specify the geometry column there, as it is not necessarily called **geom**.
4. Name the layer in the **Layer name (prefix)** field. Since we have access to every table in the database in the SQL builder, QGIS does not bother to find out which layers are involved in the query.
5. Click on **Load now!** to get the queried rows as a layer in QGIS:

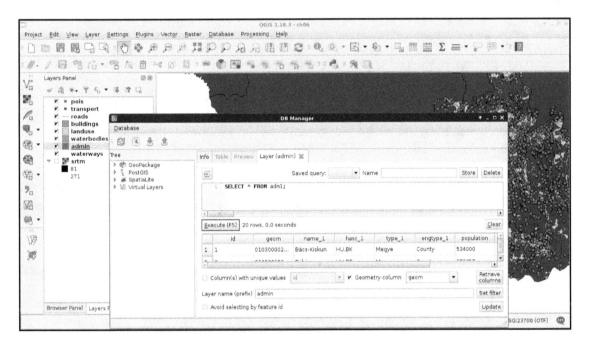

If we click on **Execute**, QGIS will only query the PostGIS database and fill the SQL window's table with the matching rows. It is great for creating quick previews without adding them as new layers in QGIS.

One of the best things in query layers is that we can modify the query on the go. If we close the SQL window, we can reopen it with our query by right-clicking on the queried layer's item in the **Layers Panel** and selecting **Update Sql Layer**. Let's update our query and select only our study area from the table. My query will look like the following:

```
SELECT * FROM adm1 WHERE name_1 = 'Baranya';
```

In QGIS's database manager, closing the query with a semicolon (;) is not necessary. However, as most RDBMSs require it (for example, MySQL or PostgreSQL with pl/pgSQL), getting used to closing queries anyways is good practice.

Additionally, vector data can hold a lot of attributes from which, often, only a few are relevant for our work. We can exclude some of the attribute columns by specifying only the relevant ones in the WHERE clause. There is one mandatory column that we have to include-- the geometry column. Let's modify our query on the administrative boundaries layer to only have the geometry column and the corrected population density column of our study area. We don't even have to include the queried name_1 column as the rows get filtered on PostgreSQL's side.

```
SELECT geom, pd_correct FROM adm1 WHERE name_1 = 'Baranya';
```

If we open the attribute table of the updated layer, we can only see two columns, _uid_ and pd_correct. We don't have to bother with the _uid_ column though, as it is QGIS's internal ID added to the attribute table of the layer.

From now on, we will recreate some of the queries we did in QGIS before. For this, we have to upload our geonames_desc.csv file to our database. Let's open it in QGIS as we did before (**Add Delimited Text Layer**) and import it in PostGIS like we imported our vector layers. QGIS's database manager is a convenient tool for not only loading vector layers in a PostGIS database but also for loading regular tables. We can leave every option of the import tool on their default values.

In PostGIS, we can calculate additional columns from existing ones on the fly. We can define these additional columns with their expressions as we did with regular columns and even give them a name with the alias keyword `AS`. Let's retrieve the whole administrative boundaries layer and, additionally, query a recalculated population density column based on the population data stored in the layer:

```
SELECT *, population / (ST_Area(geom) / 1000000) AS
    density FROM adm1;
```

	uid	pd_correct	id	name_1	hasc_1	type_1	engtype_1	population	popdensity	density
1	1	64.24	1	Bács-Kiskun	HU.BK	Megye	County	534000	44.15	64.1947964...
2	2	66.67	2	Békés	HU.BE	Megye	County	376657	45.67	66.6177002...
3	3	88.48	3	Baranya	HU.BA	Megye	County	396600	61.4	88.4136290...
4	4	99.38	4	Borsod-Aba...	HU.BZ	Megye	County	718951	66.19	99.2057970...
5	5	3000.13	5	Budapest	HU.BU	Fovaros	Capital City	1696128	2026.18	2996.91908...
6	6	98.1	6	Csongrád	HU.CS	Megye	County	423751	67.59	98.0318328...
7	7	98.66	7	Fejér	HU.FE	Megye	County	428711	67.09	98.5741826...
8	8	104.46	8	Gyor-Moson...	HU.GS	Megye	County	442667	70.34	104.335957...
9	9	86.37	9	Hajdú-Bihar	HU.HB	Megye	County	545641	58.35	86.2777615...
10	10	88.95	10	Heves	HU.HE	Megye	County	319460	59.73	88.8346821...
11	11	73.45	11	Jász-Nagyku...	HU.JN	Megye	County	403622	49.82	73.3785884...
12	12	141.44	12	Komárom-E...	HU.KE	Megye	County	315036	95.29	141.274754...
13	13	83.24	13	Nógrád	HU.NO	Megye	County	213030	55.67	83.1210006...

admin :: Features total: 20, filtered: 20, selected: 0

Show All Features

Great job! You just used your first PostGIS function, `ST_Area`, which returns the area of the given geometry. As we are in our local projection, we don't have to fear great distortions caused by a wrongly chosen projection. If we look at the attribute table, we can see some minor differences between the new `density` and the old `pd_correct` columns. That difference is due to the calculation of `pd_correct` on the surface of the WGS84 ellipsoid.

We can also do spatial queries in PostGIS. As we have features clipped to our study area, let's do something else. Select every POI inside the land use shapes. It shouldn't matter which land use shape contains the POIs, just select them all. We can do such a query with the following expression:

```
SELECT p.* FROM pois p, landuse l WHERE
    ST_Intersects(p.geom, l.geom);
```

As we used two tables for a single query, we have to exactly define which table should be included in the results. Additionally, we can ease our work by giving a shorthand for our tables in the FROM part. We can do such a thing by adding the shorthand after the table name separated by a whitespace. Let's spice up the query a little bit. Select only those POIs which are in forest areas. If we think it through, we can achieve this by filtering the geometries of the land use layer. We can include a subquery doing just that. Modify the query as follows and click on **Execute** (do not load the results as the updated layer):

```
SELECT p.* FROM pois p, landuse l WHERE ST_Intersects(
    p.geom, (SELECT l.geom WHERE l.fclass = 'forest'));
```

 Writing the shorthand separated with a whitespace is a special case of aliasing, allowed only where the intent is clear for the interpreter. If you like consistency over simplicity, you can write SELECT p.* FROM pois AS p, landuse AS l WHERE ST_Intersects(p.geom, l.geom);.

How long did it take for you? For me, it took about 80 seconds. Can you imagine QGIS loading for 80 seconds when you pan the map or zoom around? Me neither. Using subqueries in PostGIS is generally a bad way to solve problems achievable with simple queries. When we applied a filter on the land use layer and the results were correct, the filter was recalculated for every row PostGIS iterated through. PostgreSQL had no way to optimize this query and as a result it took very long. The correct way of doing this is by using a logical operator as follows:

```
SELECT p.* FROM pois p, landuse l WHERE ST_Intersects(
    p.geom, l.geom) AND l.fclass = 'forest';
```

That's more like it. But why did it work? Because PostGIS does not make spatial queries, it creates spatial joins and applies filtering. If we alter the query to include fields from the land use table, we can see what's happening. We get every property of the intersecting land use features joined to the POIs (even their geometries). This is a classic example of an inner join:

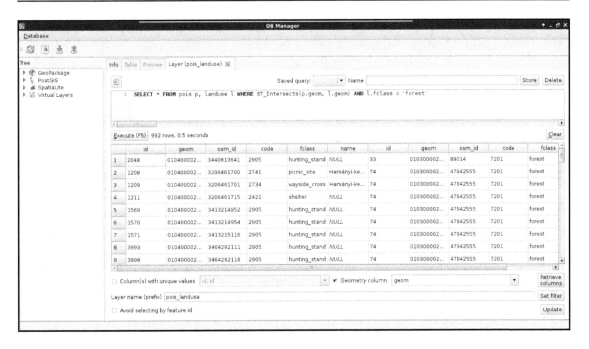

So, there we are. Joins and spatial joins. As we saw, we can do spatial joins fairly easily-- we just have to express the spatial predict with the appropriate PostGIS function in the WHERE clause. Of course, we can also use traditional join types both in regular and spatial joins. Let's create a regular join by joining the description table to our GeoNames layer. We discussed in a previous chapter that a join in QGIS is like a left outer join in an RDBMS. Therefore, we can use LEFT OUTER JOIN in our query to create similar results.

```
SELECT g.*, gd.description
  FROM geonames g
  LEFT OUTER JOIN geonames_desc gd
  ON g.featurecod = gd.code;
```

You can use a single * in the SELECT clause. That query returns correct results containing every column from the two tables both with QGIS's **Execute** button and pgAdmin's SQL builder. The only problem is that QGIS cannot load the layer if it is queried that way as some of the columns have the same name.

Let's try a spatial join as the next task. Remember when we joined our filtered GeoNames layer to our administrative boundaries layer just to have population data in our polygon layer? We can reproduce those results with the following expression:

```
SELECT a.*, g.population AS g_population
  FROM adm1 a
  INNER JOIN geonames g
  ON ST_Intersects(a.geom, g.geom) AND g.featurecod = 'ADM1';
```

There are two significant differences in this case. First, we only have an extract of our GeoNames layer containing points in our study area. Secondly, we already have a `population` column in our administrative boundaries layer. To fix the collision and make the layer readable by QGIS, we can give the `population` column of our GeoNames table an alias. As you can see, the inner join returned only the intersecting rows; therefore, we only got our study area back, the rest of the administrative boundaries layer got filtered out. We can get the whole layer, filled with data only where it is possible with the outer join, as stated before:

```
SELECT a.*, g.population AS g_population
  FROM adm1 a
  LEFT OUTER JOIN geonames g
  ON ST_Intersects(a.geom, g.geom) AND g.featurecod = 'ADM1';
```

Now we have every feature with a `g_population` column filled with `NULL` values where there is no matching GeoNames feature. Of course, we have some other join types we can use. If our target table is in the `SELECT` clause and the joined table is in the `JOIN` clause, we can make the following joins in PostgreSQL:

- `CROSS JOIN`: Creates rows for every possible combination between the two tables. It is rarely usable for spatial queries. It does not need a join condition.
- `INNER JOIN`: Returns rows from the target table where the join condition is true. The joined table's columns are only joined there.
- `LEFT OUTER JOIN`: Returns every row from the target table. Where the join condition is met, the values of the joined table are included. Where not, the fields are filled with `NULL` values.
- `RIGHT OUTER JOIN`: Returns every row from the joined table. Where the join condition is met, the values of the target table are included. Where not, the fields are filled with `NULL` values.
- `FULL OUTER JOIN`: Returns every row from both the tables. There will be only completely filled rows where the join condition is true. Other rows are partially filled with `NULL` values. It cannot be used with spatial conditions.

You can use LEFT, RIGHT, and FULL without specifying OUTER as they can only qualify an outer join; therefore, PostgreSQL automatically assumes it is an outer join. If you use simply JOIN, it is assumed you would like to create an INNER JOIN.

Summary

In this chapter, we learned a lot about spatial databases. We learned how to create a PostGIS database easily and directly from QGIS. We also learned how to query those PostGIS tables and create visualizations in QGIS from them. We used basic queries, spatial queries, regular joins, and spatial joins to achieve our goals.

In the next chapter, we will dive deeper into PostgreSQL's and PostGIS's structure. We will learn about features easily accessible from pgAdmin. These features can make our work easier and our database more manageable.

7
A PostGIS Overview

In the previous chapter, we got introduced to the various types of spatial databases. We created and filled a PostGIS database with vector and raster layers. After that, we learned about the PostgreSQL SQL syntax, and executed some basic queries to get results which were previously only possible with geoalgorithms in QGIS. With our current knowledge, we would be able to integrate PostGIS into our workflow, and create some spatial analysis and visualization tasks using QGIS only as a thin client. That means, PostGIS does the hard lifting, while we only visualize the results in QGIS. However, we are yet to explain how to create good spatial databases with the intent of creating a distributed working environment. In this chapter, we will learn about some of PostgreSQL's and PostGIS's features, which can aid us in creating a stable and well-organized spatial database.

We will cover the following topics in this chapter :

- PostgreSQL features
- PostGIS structure
- Optimizing queries
- Backing up data

Customizing the database

Most of the activities in this chapter will take place in the pgAdmin environment. We will learn how to utilize the convenient functionality it offers to create a nice, well-structured database without thinking of long commands. Of course, GUI operations cannot be automated, and using the CLI is generally faster (if we know exactly how to use it); therefore, we will also see those commands, as pgAdmin actually builds them from the options we include, and lets us see the result.

First of all, let's see what pgAdmin has to show us if we open our database:

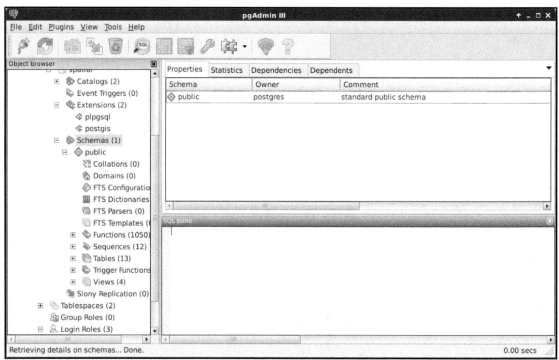

 Don't worry if you use pgAdmin 4--it has more items to show. However, the ones we will use are present in both versions.

As shown in the preceding screenshot, we can see the hierarchical structure of PostgreSQL having extensions, schemas which contain tables and functions, and roles.

Securing our database

So far, we used the `public` schema with the `postgres` role to create and fill our database. These are the default values, which are great for creating a local database, but far from ideal if we would like to create a remote GIS server. Let's discuss how roles and schemas work in PostgreSQL.

Roles are basically users that can log in, and do some stuff based on their permissions. Different roles can have different access levels to different databases. There are two kinds of roles--login roles and no-login roles (group roles). Group roles act as groups in operating systems; therefore, they can group multiple login roles, and manage their permissions in one place. Login roles are the typical users with passwords. The catch is, roles and group roles are independent from users and groups used by the operating system. System users cannot log in the database with their usernames, if a role was not created for them. Following this analogy, new users do not get roles created for them automatically by PostgreSQL.

Roles can have individual permissions, but they can also be managed by a group role. Similar to traditional users, roles can also have superuser (admin) capabilities. These superusers can modify roles and databases; therefore, they are quite dangerous to use as regular roles in a remote server. Additionally, they bypass every permission check, which makes them even more dangerous if exposed. The role `postgres` is a superuser role, which cannot be modified or dropped. It is completely fine to use in a local environment, as, by default, PostgreSQL does not accept connection requests from remote places, but only from the machine it is installed on.

It is still better to use a regular role in a local environment, as with a superuser role, you can accidentally overwrite sensitive values, drop tables, or, in the worst case, drop the entire database. You can also create a non-superuser role with privileges to modify databases and roles, and use the `postgres` role only if absolutely necessary.

Let's create two new login roles. One will be a regular GIS role with every kind of access to the GIS tables, while the other one will be a public role, which can only query the tables. It cannot modify them in any way; therefore, it will be safe to use by GeoServer, for example. We can create new roles by right-clicking on **Login Roles**, and selecting **New Login Role**. We must provide a name, and we should also add a password. As we do not want to create a superuser (having the `postgres` role is enough for that purpose), that's all we have to define. If we take a look at the last tab (**SQL**) after defining the required parameters, we can see the command that pgAdmin will use to create the role, which is as follows:

```
CREATE ROLE gis LOGIN ENCRYPTED PASSWORD
'md53929f8e603334cb5a8c5a632bcc3f3ac'
 VALID UNTIL 'infinity';
```

That is, it creates a role with `CREATE ROLE`, declares it as a login role with `LOGIN`, and, as it already calculated the hash of the password, it stores it directly with `ENCRYPTED PASSWORD`. It also includes a validity extent of infinity, which is superfluous.

If we wish to add a login role manually, we can simplify the command to the following:

```
CREATE ROLE gis LOGIN PASSWORD 'mypassword';
```

In this case, the password is provided in plain text, and the hash is calculated by PostgreSQL. We can try this method out by creating the other role from the command line by opening an SQL window (**Execute arbitrary SQL queries** in pgAdmin 3 and **Tools | Query Tool** in pgAdmin 4). If the tool is disabled, first select a database by clicking on it. If we run the following query, we should be able to see the new role created:

```
CREATE ROLE pubgis LOGIN PASSWORD 'pubpass';
```

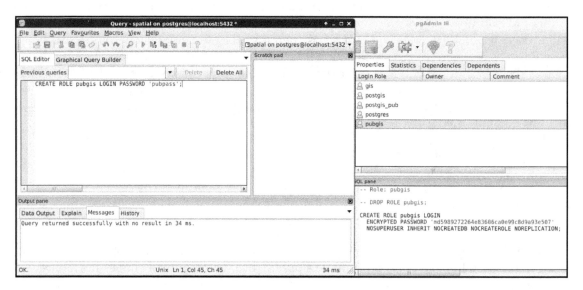

 Although MD5 is not considered a secure encryption method, by using a salt, PostgreSQL does a fair job in most of the cases. If you want access to better encryption algorithms, like bcrypt, you can use the `pgcrypto` extension; however, you also have to set up your own authentication system. You can learn more about pgcrypto at `https://www.postgresql.org/docs/9.5/static/pgcrypto.html`.

Now that we have some roles, we can give them privileges to administer or just query tables. However, doing so is very cumbersome, as these kinds of privileges do not apply on tables created later on. To solve this issue, PostgreSQL uses schemas to group tables and a lot of other things. In PostgreSQL, schemas are similar to folders in a file system. A database groups different schemas, while a single schema groups different tables. Schemas are like group roles for user management. The only difference is that using a schema is mandatory.

There are three different schemas--`information_schema`, `pg_catalog`, and `public`. The first two are system schemas used by the RDBMS internally, and showed as **Catalogs** in pgAdmin.

We should not alter those schemas in any way. The last is the default one in which we stored our data and PostGIS functionality. Having multiple schemas is a convenient way of organizing a big database, where only a single part contains spatial data. If we select the `public` schema, we can see the SQL command that creates it:

```
CREATE SCHEMA public
  AUTHORIZATION postgres;

GRANT ALL ON SCHEMA public TO postgres;
GRANT ALL ON SCHEMA public TO public;
COMMENT ON SCHEMA public
  IS 'standard public schema';
```

Apart from the comment section, every command is important when creating a schema. We have to create it with `CREATE SCHEMA`, assign an owner with `AUTHORIZATION`, and give privileges to roles with the `GRANT` expressions. By using `GRANT ALL`, PostgreSQL automatically gives every privilege to `postgres` and `public`. The only problem is that we don't have a public role. By giving every permission to `public`, PostgreSQL implicitly says that every role ever created in this database should have every privilege to this schema. Sounds dangerous? It is very convenient though if used carefully.

 You can revoke granted permissions with the `REVOKE` expression. If you would like to use the public schema in a safe way, you can use `REVOKE ALL ON SCHEMA public FROM public;` to revoke implicit privileges.

As a rule of thumb, we shouldn't keep any tables in the `public` schema which shouldn't be accessed and edited by every role. Let's put our spatial database in a dedicated schema with the following steps:

1. Create a new schema by right-clicking on **Schemas** and selecting **New Schema**.
2. Give a name to the new schema, and assign an owner. The owner should be the role we created for managing the spatial database (for me it is **gis**).
3. Click on **OK** to create the schema.

4. Open a new SQL window, and grant some schema privileges to the roles we created.

   - Grant every privilege to the administrator role by running the expression GRANT ALL ON SCHEMA spatial TO gis;.
   - Grant only select privileges to the public role by running the expression GRANT USAGE ON SCHEMA spatial TO pubgis;.

5. In the SQL window, also grant some table privileges to the roles.

   - Grant every privilege to the administrator role with ALTER DEFAULT PRIVILEGES IN SCHEMA spatial GRANT ALL ON TABLES TO gis;.
   - Grant only select privileges to the public role with ALTER DEFAULT PRIVILEGES IN SCHEMA spatial GRANT SELECT ON TABLES TO pubgis;.

6. Move every table to the new schema. Right-click on the tables in the public schema, and select **Properties**. There we can alter these properties. Watch out not to move the spatial_ref_sys table created by PostGIS. We can optionally set the owner of the tables to our GIS role:

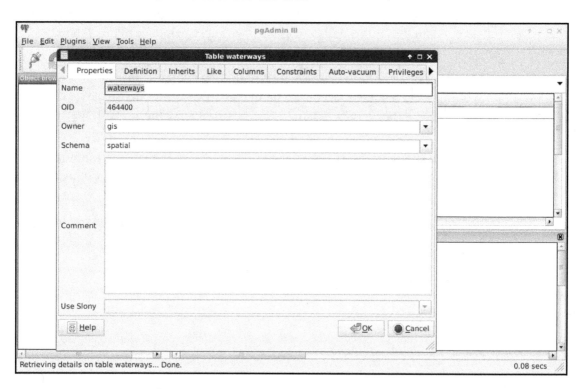

Schema and table privileges differ; therefore, if you would like to fine-tune the privilege system, you cannot avoid using table permissions. On schemas, you can grant USAGE, CREATE, and ALL. For the rest of the privileges, you can read the PostgreSQL documentation at https://www.p ostgresql.org/docs/9.4/static/sql-grant.html. Make sure you select the appropriate documentation version.

We are all set. Let's try out our new schema by doing a spatial query that we've already done before:

```
SELECT g.* FROM landuse l, geonames g
WHERE ST_Intersects(l.geom, g.geom);
```

The preceding query returns an error. As we moved our tables out of the public schema, we have to explicitly define the schema of the tables with the syntax schema.table. If we update our query appropriately, PostGIS returns the intersecting features:

```
SELECT g.* FROM spatial.landuse l, spatial.geonames g
WHERE ST_Intersects(l.geom, g.geom);
```

I'm completely sure you've already found out why PostGIS still works. As we left it in the public schema, its functionality remained exposed. Can we move it out to our new schema, and use its functions by prefixing them with the schema name? Not easily. PostGIS quite heavily relies on sitting in the public (or a similarly exposed) schema. However, as we witnessed, we can move out our spatial tables wherever we see them fit.

As we stated before, using the public schema is quite convenient. One of the reasons for this is that its content can be accessed directly without specifying the schema name. PostgreSQL achieves this by using a search path in which the public schema is defined. You can query the search path variable with the SHOW search_path; expression, and modify it with the SET search_path TO 'newsearchpath'; command. Adding additional schemas will make their content also available without prefixing. Be careful with this approach, though. Schemas contain local objects by design; thus, different schemas can share the same object (like table, function, and so on) names.

As the final task, let's open QGIS, and edit our PostGIS connection to use our new **gis** role. We have to use the database manager to get our layers again, as the project file for the previous chapter still thinks they are placed in the `public` schema, and, therefore, it cannot access them:

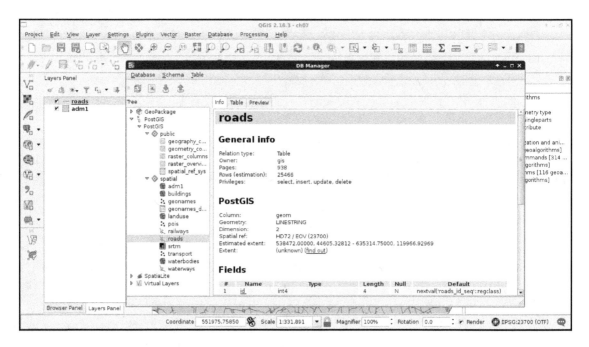

Constraining tables

From the database and its schemas, we arrive at tables. As we know, databases can hold multiple schemas, while a single schema can hold multiple tables. Tables have typed columns and items as rows. What we did not discuss before is that tables can have a lot more properties. We can consider these constraints, rules, triggers, and indexes metadata, and we can set them with expressions, or from pgAdmin. If we inspect one of our tables, we can see the definition it was created with, which is as follows:

```
CREATE TABLE spatial.adm1
(
 id serial NOT NULL,
 geom geometry(Polygon,23700),
 name_1 character varying(75),
 hasc_1 character varying(15),
 type_1 character varying(50),
 engtype_1 character varying(50),
```

```
   population integer,
   popdensity double precision,
   pd_correct double precision,
   CONSTRAINT adm1_pkey PRIMARY KEY (id)
)
WITH (
   OIDS=FALSE
);
```

If you've used an RDBMS before, this verbose CREATE TABLE expression resembles the usual ones; however, some of the lines are different. On the other hand, if we try to use a more regular expression, we end up with a similar table:

```
CREATE TABLE spatial.test (
   id serial PRIMARY KEY NOT NULL,
   geom geometry(Point,23700),
   attr varchar(50)
);
```

The first difference is that, in RDBMSs, keys are realized as constraints in the database. It is not just for keys, though. Almost every qualifier (except NOT NULL) is realized as a constraint. The most important qualifiers for tuning a database are NOT NULL, UNIQUE, PRIMARY KEY, and FOREIGN KEY. By adding the NOT NULL definition to a column, we explicitly tell PostgreSQL not to accept updates or new rows with an empty value for that field. Setting it for most of the columns is a good practice, as it can help to make the table more consistent. For example, it will prevent everyone from adding new features or updating existing ones without supplying the required attributes.

There are two kind of keys--primary key and foreign key. They are used to logically link tables together for various purposes. For example, we can fight redundancy by creating multiple tables, and linking them together with keys. They are also useful for creating a cascading structure and defining rules for what should happen to the referenced row when the other one changes. This is what really makes relational databases relational.

If we alter the spatial.test table created before with some updated definitions, we end up with two named constraints:

```
ALTER TABLE spatial.test
ADD UNIQUE (attr),
ALTER COLUMN attr SET NOT NULL;
```

So far, we can see that PostgreSQL queries work in the concise way we might be used to. However, if we inspect the table in pgAdmin, it crafts more verbose, more explicit expressions to achieve the same results. We can also alter our tables in pgAdmin. We have to open the **Properties** of a table to reach the relevant options under the **Columns** and **Constraints** tabs. Under **Columns**, we can select any column, click on **Change**, and check the **Not NULL** checkbox in the **Definition** tab. For setting the unique constraint, we have to add a new, **Unique** item in the **Constraints** tab. In the dialog, we only have to provide the column name in the **Column** tab, and click on **Add**:

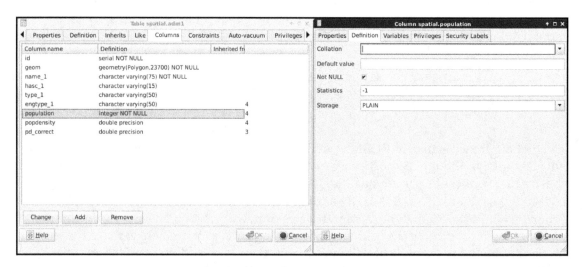

 While adding a UNIQUE definition would theoretically prevent PostGIS from storing duplicate geometries, it is neither feasible, nor often possible to apply it to the geometry column, due to the size of the contained geometries.

The last constraint we should explore is CHECK. We can create rules that the columns have to comply with before an update or insert occurs by using regular SQL expressions. For example, we can create a CHECK for the population column, and use the population > 0 expression to enforce a positive population value. Another reason to use a CHECK constraint is to validate the geometries themselves, as PostGIS only checks that they are using the right geometry type. For this task, we have to create a CHECK, as we would add a UNIQUE definition:

1. In one of the spatial tables' **Properties** window, select the **Constraints** tab.
2. Add a new **Check**.
3. In the **Definition** tab, supply the following expression--ST_IsValid(geom):

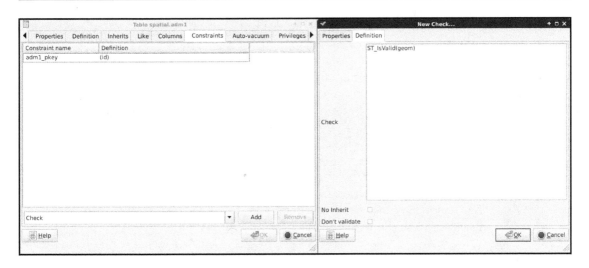

If we look at the SQL tab, we can see that defining a CHECK is very similar to defining other constraints with an expression:

```
ALTER TABLE spatial.adm1
  ADD CHECK (ST_IsValid(geom));
```

Let's see a real world example. PostGIS has the capability to handle curves. Besides the regular points, lines, polygons, and their multipart counterparts, PostGIS handles the following geometry types:

- **Triangle**: A simple triangle
- **Circular String**: A basic curve type, which describes a circular arc with two end points and minimum of one additional point on the circle's perimeter
- **Compound Curve**: A geometry mixing line strings and circular strings
- **Multi Curve**: The multipart geometry mixing line strings, circular strings, and compound curves
- **Curve Polygon**: A polygon-like geometry, although the polygon can be made of line strings, circular strings, and compound curves
- **Multi Surface**: A multipart geometry mixing regular polygons and curve polygons
- **Polyhedral Surface**: A surface made of polygons--useful for storing 3D models.
- **TIN (Triangulated Irregular Network)**: A surface made of triangles--useful for representing **Digital Elevation Models** (**DEM**) with vector data

Sometimes, we would like to store curves to have a representation which can be visualized instantly without post-processing. Unfortunately, there is no function in PostGIS which smoothes lines with a nice smoothing algorithm. To fill this gap, I created a small and primitive script, which converts corners to curves based on some rough approximations. First of all, let's get this script (`createcurves.sql`) from the supplementary material's `ch07` folder, or download directly from `https://gaborfarkas.github.io/practical_gis /ch07/createcurves.sql`. We can open the file in pgAdmin's SQL window, and run the content as a regular SQL expression. When we are done, we should have access to the `CreateCurve` function.

You can use this function for any purpose; however, it is neither mathematically, nor computationally, optimal. Do not trust the results, always review them before using.

In this example, we will build a table storing the curvy representation of our rivers table. First of all, we need to create a table which will contain our curves. We could save the results of a query directly into a table; however, we will also apply some logic, which is as follows:

- Only the IDs and the geometries get stored.
- PostGIS must only accept compound curves as geometries, as our custom function can only return compound curves when fed with line strings.
- The IDs reference the IDs of the original table. If we delete a row from the original waterways table, PostgreSQL must also delete its curvy representation.
- If we update the original table, PostgreSQL must automatically update the curve table accordingly.

We can apply some of this logic when we create the table with the correct table definition in an SQL window as follows:

```
CREATE TABLE spatial.waterways_curve (
  id integer PRIMARY KEY UNIQUE REFERENCES spatial.waterways (id)
  ON DELETE CASCADE,
  geom geometry(CompoundCurve, 23700)
);
```

With this preceding definition, we defined two columns: one for the IDs, and one for the geometries. The IDs are foreign keys, as they reference another column in another table. By adding the `ON DELETE CASCADE` definition, we ask PostgreSQL to delete referencing rows from this table when rows get deleted from the parent table.

The geometry is defined as compound curves in our local projection (the number should reference your local SRID). This table definition alone fulfilled some of our requests. There is a great feature in PostgreSQL which we can apply to the rest of them--rules.

Before creating the rules, we should synchronize our new table with the waterways table. We can insert values into an existing table with the INSERT INTO expression. What is more exciting is that we can shape a SELECT statement to insert the results instantly into a destination table as follows:

```
INSERT INTO spatial.waterways_curve (
  SELECT id, CreateCurve(geom) FROM spatial.waterways
);
```

Rules are custom features of PostgreSQL, which allow us to define some logic when we select, update, delete, or insert rows in a table. In a rule, we can define if we would like to apply our custom logic besides the original query, or instead of it. In order to define a rule, we can select our waterways table in pgAdmin, right-click on Rules, and select **New Rule**:

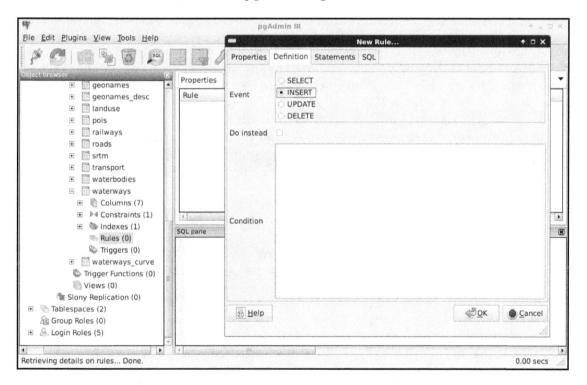

The rule maker of pgAdmin is a little bit scattered, as rules are quite complex. There are these three distinct tabs we need to fill in order to have a rule:

- The first thing we can define is the rule's name in the **Properties** tab (for example, `waterways_insert`).
- In the **Definition** tab, we can define the event we would like to hook our rule onto. Rules are like event listeners in object-oriented languages. They are invoked when an event we hooked them onto occurs. We should select **INSERT** for our first rule. We can also supply a condition, which has to happen in order to run the rule. We can leave that field empty.
- The **Statements** tab holds the body of our rule. Every logic we would like to execute when our event triggers goes there. As we need PostgreSQL to automatically generate curves from new geometries, we can insert some of the new row's values into the curve table directly:

```
INSERT INTO spatial.waterways VALUES (
  NEW.id, CreateCurve(NEW.geom)
);
```

As we are now inserting literal values directly into a table, we have to provide the `VALUES` keyword. Other RDBMSs might also require us to provide the columns we would like to insert our values into; however, PostgreSQL is smart enough to find out that we only have two columns in the new table with matching types. There is one additional thing we need to keep in mind when writing rules and triggers--row variables. We can access the processed rows with the keywords `NEW` and `OLD`. Of course, we should only use the appropriate one from the two. That is, if we write a rule for an `INSERT`, we should use `NEW`. If we write a rule for an `UPDATE`, we can use both of them, while for `DELETE` events, we should only use `OLD`.

You can only create rules if you own the table. Keep in mind that if you are still connected in pgAdmin with the `postgres` role, your new tables' owner becomes `postgres`. Therefore, you cannot create rules for those tables with other roles. You can still change the ownership of the tables to do that. Also note that, as superusers are excluded from permission checks, `postgres` can create rules to every table in the database.

The second rule should occur when we update the `waterways` table. In this case, we should only update the curve table when a geometry changes; therefore, we need a condition besides the rule body:

1. Name the rule in the **Properties** tab.

2. Select the **UPDATE** event in the **Definition** tab, and add the following expression as a Condition--`NOT ST_Equals(OLD.geom, NEW.geom)`.

3. Write the rule's body in the **Statements** tab as follows:

```
UPDATE spatial.waterways_curve SET geom = CreateCurve(NEW.geom)
  WHERE id = NEW.id;
```

As you can see, we can directly compare the new and old values of the updated row, but PostgreSQL has no idea how it should update the curve table. Therefore, we have to supply a `WHERE` clause defining where we would like to update our geometry if it differs from the old one. Of course, we can also use `OLD.id`, as the two should be the same.

We don't have to create a rule for deleting rows, as the cascading constraint on the foreign key will make sure that the rows deleted from the waterways table get deleted from the curve table. Let's see, instead, what we created in QGIS. First of all, we can see in the database manager that QGIS recognizes the compound curves we have as line geometries, as it also supports circular strings. If we open the layer, and zoom in to a corner, we can see our curves:

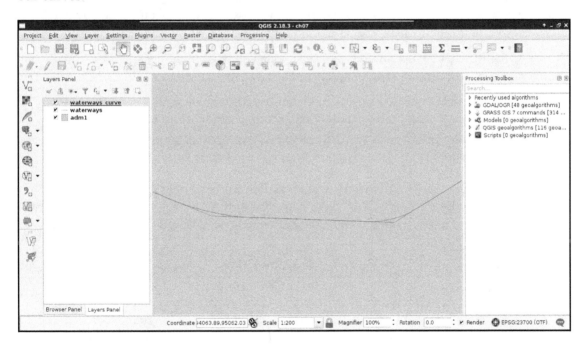

To check if everything works as expected, we have to edit our waterways table a little bit, in the following way:

1. Select the `waterways` layer in the **Layers Panel**.
2. Start an edit session with the pencil-shaped **Toggle Editing** button in the main toolbar.
3. Select the **Node Tool** from the freshly enabled tools in the main toolbar.
4. Click on one of the lines, and move one of its vertices a little bit with the red anchor.
5. Save the edit by clicking on the **Save Layer Edits** button.
6. Refresh the layers by panning, zooming, or clicking on the **Refresh** button in the main toolbar.

If everything is configured properly, we should be able to see our curve layer following the edits we made in the waterways layer. The only problem left is that we can update the curve table manually, getting the tables out of synchronization. We can easily avoid this, however, by smart permission control. What do we know about rules? Only table owners can define them on tables; therefore, PostgreSQL will run them on behalf of the owner. Thus, we can fine-tune the privilege system in the following way:

1. Set the curve table's owner to a superuser role, like `postgres`.
2. Set the waterways table's owner to a superuser role, like `postgres`.
3. Give select privilege to the GIS role on the curve table with the expression GRANT SELECT ON TABLE spatial.waterways_curve TO gis;.
4. Give every privilege to the GIS role on the waterways table with the expression GRANT ALL ON TABLE spatial.waterways TO gis;.

Now we can modify the waterways table with the GIS role, but only the `postgres` superuser can modify the curve table. That is, we can only access the curves layer with the GIS role, but the rules we defined change the curve table accordingly and automatically when we modify the waterways table.

 Note that using rules increases stability and consistency, but also degrades performance. Most of the rules can be rephrased as triggers, which are often faster, but they need to be declared as functions in a programming language usable by PostgreSQL. The most SQL-like language (besides SQL) for this is PostgreSQL's PL/SQL extension, pl/pgSQL. You can see an example if you inspect the `CreateCurve` function's source.

Saving queries

As we've already witnessed, using PostGIS has a great advantage of offering flexible results over traditional desktop GIS applications. That is, we can play with queries without filling the memory or disk with useless intermediate data showing wrong results. But how can we save the correct results once we've found out the right query to produce it? There are various ways of saving results in PostgreSQL. The most basic way is to save them right into a new table. All we have to do is to prefix our query with the `CREATE TABLE tablename AS` expression.

Let's try it out by creating another curve table with the following expression:

```
CREATE TABLE spatial.tempcurve AS SELECT id,
  CreateCurve(geom) AS geom FROM spatial.waterways;
```

If we refresh the tables, or import the `tempcurve` table in QGIS, the geometries are the same as in our fine-tuned `waterways_curve` table. With these kinds of queries, PostgreSQL creates a regular table, finds out the column types from the queried columns, and fills this new table with the query results. The only differences from the PostgreSQL perspective in the two tables are the constraints and rules we added, which can be also defined on an existing table.

On the other hand, there is an important PostGIS difference between the two methods. PostgreSQL cannot find out the type of the geometries we have, therefore, it types the geometry column simply as `geometry`. That means, we lose the subtype information, and end up with a column which does not care for geometrical consistency. On top of that, it doesn't even care for the projection we use. Luckily, there's a method for telling PostgreSQL the subtype we would like to use--explicit typing. If we cast the results of `CreateCurve` to a compound curve geometry in our local projection, PostgreSQL can safely use our preferred subtype:

```
CREATE TABLE spatial.tempcurve AS SELECT id,
  CreateCurve(geom)::geometry(CompoundCurve,23700) AS geom
  FROM spatial.waterways;
```

Don't forget to use your local projection, or the projection your waterways table is in, instead of my **EPSG:23700**. Furthermore, make sure you drop the `tempcurve` table by right-clicking on it, and selecting **Delete/Drop** before running the query again.

This method is very useful for storing quickly accessible versions of our results, but the tables we create this way remain static. Our heroic attempt at synchronizing the results with the data source was, of course, a very nice way to get rid of this obstacle. However, this is not always a practical method due to the hassle it involves. To create dynamic results, which change with the data source, we can build views. In PostgreSQL, views are special empty tables, which have a rule hooked on to their **SELECT** event. That rule simply executes the query we saved our view with. As a result, we can save our query in a view, which means that it gets executed every time we access the view. If we look at our public schema, we can see the views that PostGIS created. They dynamically query the database, and create catalogues of our data in it with lengthy and complex expressions. Let's create our own view the same way we created a table, as follows:

```
CREATE VIEW spatial.tempcurve AS SELECT id,
  CreateCurve(geom)::geometry(CompoundCurve,23700) AS geom
  FROM spatial.waterways;
```

In QGIS, we can see that the new view is recognized as a view with its definition. However, if we load the layer, we also stumble on to the performance cut it introduced. As views are basically saved queries, which are executed every time the canvas is refreshed (for example, on panning and zooming), and the `CreateCurve` function is slow, storing this table as a view has a great performance impact:

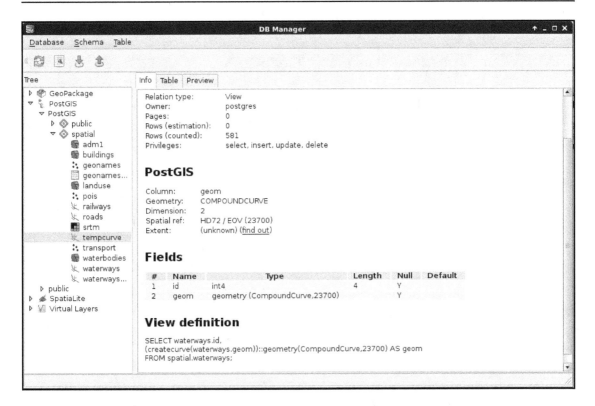

On the other hand, those PostGIS catalogues are only queried once in a while, and it is completely affordable to sacrifice some speed to have dynamic tables without the extra hassle. On those PostGIS views, we can identify some rules for inserting, updating, and deleting rows. They are needed, as views are generally modifiable. If we try to update, insert into, or delete from a view, PostgreSQL tries to find out which tables are affected, and applies the required operations on them. To override this behavior, we can define three simple rules on views using the following scheme:

```
CREATE OR REPLACE RULE tablename_event AS
 ON event TO tablename DO INSTEAD NOTHING;
```

You can see a default **_RETURN** rule in every view. This is the **SELECT** rule created by PostgreSQL by default, returning the results of the underlying query.

We can see an example of such rules on the following screenshot:

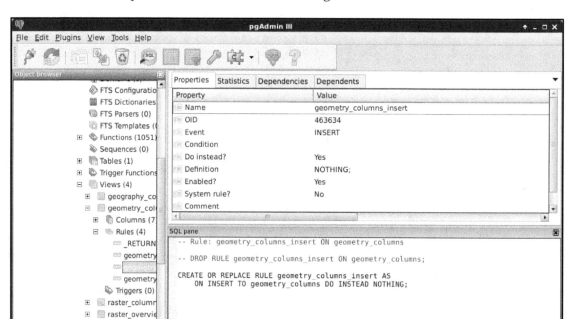

The `event` should be `UPDATE`, `INSERT`, or `DELETE`, while the `tablename` should be the view we would like to apply the rules on. Of course, we can do it with pgAdmin's GUI like we did before. In there, we only have to give the rule a name, select the event type, and check the **Do instead** checkbox. If we leave everything else blank, the simple rule given earlier gets created. By applying these rules on the three events, we can easily make our views read-only. If anyone would like to modify our tables via a protected view, those operation requests will simply bounce off the database.

What if we would like to create views with better performance? We shouldn't be so demanding, right? Well, PostgreSQL thinks otherwise, and happily offers us materialized views. These views store snapshots of the queries stored in them. We can create such a view by running the following query:

```
CREATE MATERIALIZED VIEW spatial.tempcurve AS SELECT id,
  CreateCurve(geom)::geometry(CompoundCurve,23700) AS geom
  FROM spatial.waterways;
```

As a result, we get a view which has data in it, and is read-only by default. We cannot insert into it, update it, or remove rows from it. The only drawback is that it stores the snapshot of the query created at the time of execution. If we would like to incorporate changes in our materialized view, we have to refresh it manually as follows:

```
REFRESH MATERIALIZED VIEW spatial.tempcurve;
```

You can right-click on a materialized view in pgAdmin, and select **Refresh data** to refresh it with the GUI.

Optimizing queries

We already used some techniques to speed up queries, although we have a lot more possibilities in tuning queries for faster results. To get the most out of our database, let's see how queries work. First of all, in RDBMS jargon, tables, views, and other similar data structures are called relations. Relation data is stored in files consisting of static-sized blocks (pages). In PostgreSQL, each page takes 8 KB of disk space by default. These pages contain rows (also called tuples). A page can hold multiple tuples, but a tuple cannot span multiple pages. These tuples can be stored quite randomly through different pages; therefore, PostgreSQL has to scan through every page if we do not optimize the given relation. One of the optimization methods is using indexes on columns just like the geometry columns, other data types can be also indexed using the appropriate data structure.

Don't worry about storing values larger than 8 KB. Complex geometries can easily exceed that size. PostgreSQL stores large values in external files using a technique, which is called by the awesome acronym **TOAST (The Oversized-Attribute Storing Technique)**.

There is another optimization method which can speed up sequential scans. We can sort our records based on an index we already have on the table. That is, if we have queries that need to select rows sequentially (for example, filtering based on a value), we can optimize these by using a column index to pre-sort the table data on disk. This causes matching rows to be stored in adjacent pages, thereby reducing the amount of reading from the disk. We can sort our waterways table with the following expression:

```
CLUSTER spatial.waterways USING sidx_waterways_geom;
```

In this preceding query, the `sidx_waterways_geom` value is the spatial index's name, which was automatically created by QGIS when we created the table. Of course, we can sort our tables using any index we create.

 Sorting a table only works while we do not modify the rows' order. PostgreSQL will not care about keeping the right order if we remove, insert, or update rows. On the other hand, once we use `CLUSTER` to sort a table, the following ordering operations won't need the index to be specified. For example, we can order our **waterways** table with the expression `CLUSTER spatial.waterways;`.

For the next set of tuning tips, we should understand how an SQL expression is turned into a query. In PostgreSQL, there are the following four main steps from an SQL query to the returned data:

1. **Parsing**: First, PostgreSQL parses the SQL expression we provided, and converts it to a series of C structures.
2. **Optimizing**: PostgreSQL analyzes our query, and rewrites it to a more efficient form if it can. It strives to obtain the least complex structure doing the same thing.
3. **Planning**: PostgreSQL creates a plan from the previous structure. The plan is a sequence of steps required for achieving our query with estimated costs and execution times.
4. **Execution**: Finally, PostgreSQL executes the plan, and returns the results.

For us, the most important part is planning. PostgreSQL has a lot of predefined ways to plan a query. As it strives for the best performance, it estimates the required time for a step, and chooses a sequence of steps with the least cost. But how can it estimate the cost of a step? It builds internal statistics on every column of every table, and uses them in sophisticated algorithms to estimate costs. We can see those statistics by looking at the PostgreSQL catalog's (`pg_catalog`) **pg_stats** view:

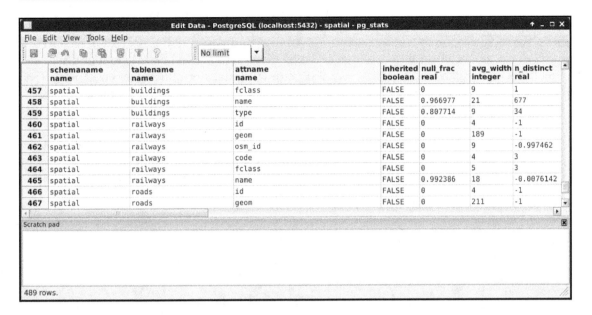

	schemaname name	tablename name	attname name	inherited boolean	null_frac real	avg_width integer	n_distinct real
457	spatial	buildings	fclass	FALSE	0	9	1
458	spatial	buildings	name	FALSE	0.966977	21	677
459	spatial	buildings	type	FALSE	0.807714	9	34
460	spatial	railways	id	FALSE	0	4	-1
461	spatial	railways	geom	FALSE	0	189	-1
462	spatial	railways	osm_id	FALSE	0	9	-0.997462
463	spatial	railways	code	FALSE	0	4	3
464	spatial	railways	fclass	FALSE	0	5	3
465	spatial	railways	name	FALSE	0.992386	18	-0.0076142
466	spatial	roads	id	FALSE	0	4	-1
467	spatial	roads	geom	FALSE	0	211	-1

Scratch pad

489 rows.

As the planner uses these precalculated statistics, it is very important to keep them up to date. While we don't modify a table, the statistics won't change. However on frequently changed tables, it is recommended not to wait on the automatic recalculations, and update the statistics manually. This update usually involves a clean-up, which we can do on our `waterways_curve` table by running the following expression:

```
VACUUM ANALYZE spatial.waterways_curve;
```

PostgreSQL does not remove the deleted rows from the disk immediately. Therefore, if we would like to free some space when we have some deleted rows, we can do it by vacuuming the table with the `VACUUM` statement. It also accepts the `ANALYZE` expression, which recalculates statistics on the table. Of course, we can use `VACUUM` without `ANALYZE` to only free up space, or `ANALYZE` without `VACUUM` to only recalculate statistics. It is just a good practice to run a complete maintenance by using them both.

Now that we know how important it is to have correct statistics, and how we can calculate them, we can move on to real query tuning. Analyzing the plan that PostgreSQL creates involves the most technical knowledge and experience. We should be able to identify the slow parts in the plan, and replace them with more optimal steps by altering the query. One of the most powerful tools of PostgreSQL writes the query plan to standard output. From the plan structured in a human readable form, we can start our analysis. We can invoke this tool by prefixing the query with EXPLAIN. This statement also accepts the ANALYZE expression, although it does not create any kind of statistics. It simply runs the query and puts out the real cost, which we can compare to the estimated cost. Remember that inefficient query from the last chapter? Let's turn it into a query plan like this:

```
EXPLAIN ANALYZE SELECT p.* FROM spatial.pois p,
    spatial.landuse l WHERE ST_Intersects(p.geom,
    (SELECT l.geom WHERE l.fclass = 'forest'));
```

We can see the query plan generated as in the following screenshot:

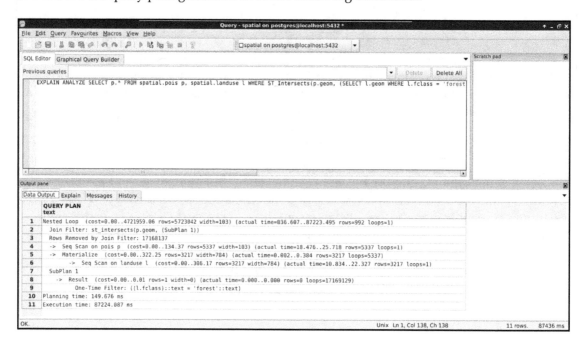

According to the query plan, in my case, we had to chew ourselves through about 5,700,000 rows. One of the problems is that this query cannot use indices, therefore, it has to read a lot of things into memory. The bigger problem is that it had to process more than 17 million rows, much more than estimated in the query plan. As PostGIS does spatial joins in such queries, and we provided a subquery as one of the arguments, PostGIS did a cross join creating every possible combination from the two tables. When I checked the row numbers in the tables, the POI table had 5,337 rows, while the land use table had 3,217 rows. If we multiply the two numbers, we get a value of 17,169,129. If we subtract the 17,168,137 rows removed by the join filter according to the executed query plan, we get the 992 relevant features. The result is correct, but we took the long way. Let's see what happens if we pull out our subquery into a virtual table, and use it in our main query:

```
EXPLAIN ANALYZE WITH forest AS (SELECT geom
 FROM spatial.landuse l WHERE l.fclass = 'forest')
 SELECT p.* FROM spatial.pois p, forest f
 WHERE ST_Intersects(p.geom, f.geom);
```

By using `WITH`, we pulled out the geometries of forests into a **CTE (Common Table Expression)** table called `forest`. With this method, PostgreSQL was able to use the spatial index on the POI table, and executed the final join and filtering on a significantly smaller number of rows in significantly less time. Finally, let's see what happens when we check the plan for the final, simple query we crafted in the previous chapter:

```
EXPLAIN ANALYZE SELECT p.* FROM spatial.pois p,
 spatial.landuse l
 WHERE ST_Intersects(p.geom, l.geom)
 AND l.fclass = 'forest';
```

The query simply filters down the features accordingly, and returns the results in a similar time. What is the lesson? We shouldn't overthink when we can express our needs with simple queries. PostgreSQL is smart enough to optimize our query and create the fastest plan.

There are some scenarios where we just cannot help PostgreSQL to create fast results. The operations we require have such a complexity that neither we, nor PostgreSQL, can optimize it. In such cases, we can modify the memory available to PostgreSQL, and speed up data processing by allowing it to store data in memory instead of writing on the disk and reading out again when needed. To modify the available memory, we have to modify the `work_mem` variable. Let's see the available memory with the following query:

```
SHOW work_mem;
```

The 4 MB RAM space seems a little low. However, if we see it from a real database perspective, it is completely reasonable. Databases are built for storing and distributing data over a network. In a usual database, there are numerous people connecting to the database, querying it, and some of them are also modifying it. The `work_mem` variable specifies how much RAM a single connection can take up. If we operate a small server with 20 GB of RAM, let's say we have 18 GB available for connections. That means, with the default 4 MB value, we can have less than 5,000 connections. For a medium-sized company operating a dynamic website from that database, allowing 5,000 people to browse its site concurrently might be even worse than optimal.

On the other hand, having a private network with a spatial database is a completely different scenario. Spatial queries are usually more complex than regular database operations, and complex analysis in PostGIS can take up a lot more memory. Let's say we have 20 people working in our private network using the same database on the same small server. In this case, we can let them have a little less than 1 GB of memory for their work. If we use the same server for other purposes, like other data processing tasks, or as an **NFS** (**Network File System**), we can still give our employees 128 MB of memory for their PostGIS related work. That is 32 times more than they would have by default, and only takes roughly 2 GB of RAM from our server.

In modern versions of PostgreSQL, we don't have to fiddle with configuration files located somewhere on our disk to change the system variables. We just have to connect to our database with a superuser role, use a convenient query to alter the configuration file, and another one for reloading it and applying the changes. To change the available memory to 128 MB, we can write the following query:

```
ALTER SYSTEM SET work_mem = '128MB';
```

Finally, to apply the changes, we can reload the configuration file with this query:

```
SELECT pg_reload_conf();
```

Backing up our data

It's great fun to work with spatial databases. However, when we use a database extensively, we can lose a lot of valuable data in case of a failure. To minimize the damage involved in a server failure, we can, and we should, create backups of our database. In PostgreSQL, there are multiple great and powerful ways to create backups.

Creating static backups

The traditional way of backing up a PostgreSQL database is to create static backups. This method uses the command-line tools `pg_dump` and `pg_restore` to create and restore the whole database, or parts of it. Of course, we do not have to use those CLI tools for backing up and restoring, as pgAdmin offers us a way to use of them via its GUI. The main advantage of using static backups is that we can save only parts of the database, like only our spatial data, or just one table. Its main disadvantage, of course, is its static nature. We have to refresh the backups manually if we would like to archive a more recent version of our database.

Let's see what happens if we back up one of our tables. We can do it by right-clicking on the table in pgAdmin and selecting **Backup**. In the dialog, we have numerous options. The most important is the file format. We can choose between four formats:

- **Plain**: The selected objects will be saved as a series of regular SQL expressions. The saved file can be restored by opening the file in an SQL window, and running the query. Choosing this format allows us to read and modify the result if we would like to port our tables to another RDBMS.
- **Custom, Tar, and Directory**: These are PostgreSQL-specific formats, which compress the data nicely, and can be used with PostgreSQL's restore tool. The custom format creates a PostgreSQL backup file, the tar format creates an archive, and the directory format creates a directory structure with compressed objects representing PostgreSQL objects.

You can read more about these formats in PostgreSQL's `pg_dump` manual at `https://www.postgresql.org/docs/9.4/static/app-pgdump.html`.

To try out the plain format, we should back up our `geonames_desc` table, as it does not have geometries; therefore, `pg_dump` will create a nicely readable output.

1. Right-click on the `geonames_desc` table, and choose **Backup**.
2. Browse the output folder, and choose a name with the `sql` extension (for example, `geonames_desc.sql`).
3. Choose **Plain** in the **Format** field.
4. In the **Dump Options #2** tab (**Dump options** in pgAdmin 4), check the **Use Column Inserts** box.
5. Click on the **Backup** button. When it is done, click on the **Done** button to close the dialog.

If we open the resulting file in a text or code editor, we can see a very well-structured set of SQL expressions creating the table, the constraints, the sequence, and inserting the data:

What PostgreSQL did not save is the schema that it should restore the table in. On the other hand, if we drop our `geonames_desc` table, and run the contents of this file in an SQL window, the table gets recreated in the spatial schema. If we inspect the start of the file, the script sets some PostgreSQL-specific variables. From the numerous variables, the following overwrites the search path for the transaction:

```
SET search_path = spatial, pg_catalog;
```

Since, in the search path, the `spatial` schema is set as first, PostgreSQL will automatically put everything in there.

 There are a lot of PostgreSQL-specific expressions in an SQL dump. If you would like to port the tables to another RDBMS, you have to identify, then modify or remove those parts.

We can also create compressed archives. Let's create a backup containing our entire **spatial** schema as follows:

1. Right-click on the `spatial` schema in pgAdmin, and select **Backup**.
2. Browse the output folder, and choose a name with the backup extension (for example, `spatial.backup`).
3. Choose **Custom** in the **Format** field.
4. Click on the **Backup** button. When it is done, click on the **Done** button to close the dialog.

One of the disadvantages of using `pg_dump` for creating backups is that it saves everything from the dumped objects. If we dump the entire database, it saves every PostGIS object (like functions) along our data. This is another good reason for using a different schema for the actual spatial data, as this way we can back up the relevant data only.

To restore the data dump, we have to specify the place where we would like to extract our archive. If we exported a table, we have to right-click on the schema we would like to insert it into, while, if we dumped a schema, we have to right-click on the database. There we have to select **Restore**. A great perk of using a compressed archive for a backup is that we can browse through the exported objects, and specify the ones we would like to restore:

1. Right-click on the database, and click on **Restore**.
2. Browse the backup archive created previously.

3. Click on **Display objects**, and browse the dumped objects in the **Objects** tab:

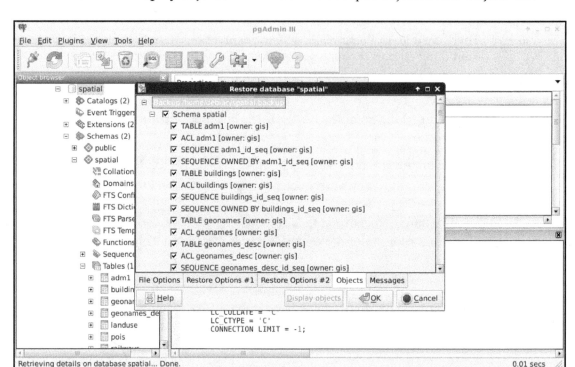

In pgAdmin 4, there is currently no way to inspect an archive and only restore a part of it.

Continuous archiving

In some setups, it is simply inconvenient to save static backups of a database on regular intervals. With continuous archiving, we can archive the changes made to our database, and roll back to a previous stable state on failure or corruption. With this archiving method, PostgreSQL automatically saves logs in a binary format to a destination location, and can restore the whole database from those logs if necessary. The main disadvantage of this method is that the whole cluster is saved, and there is no way to specify which parts we would like to archive.

First of all, what is a cluster? In PostgreSQL terms, a cluster contains every data stored in a PostgreSQL installation. A cluster can contain multiple databases containing multiple schemas with multiple tables. Using continuous archiving is crucial in production servers where corruption or data loss is a real threat, and the ability to roll back to a previous state is required.

First of all, let's find out where our PostgreSQL cluster is located on the disk. The default path is different on different operating systems, and, besides that, we can specify a custom location for our cluster. For example, as I use an SSD for the OS, and PostgreSQL would store its database on the SSD by default, I specified the `postgres` folder in a partition of my HDD mounted at `/database` for the database. We can see the path to our cluster by running the following query:

```
SHOW data_directory;
```

If we open the path we got from the previous query in a file manager, we will see the files and folders our PostgreSQL cluster consists of. From the folders located there, the `pg_xlog` contains the WALs (Write Ahead Logs) of our database transactions. WALs are part of PostgreSQL's ACID implementation, as it can restore the last stable state from these logs if something bad happens. They can be also used for continuous archiving by saving them before PostgreSQL recycles them:

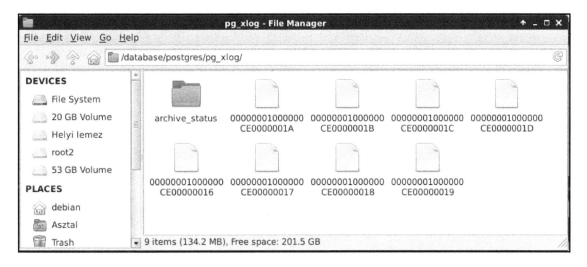

 If you cannot access the cluster with a regular user, it is completely fine. The cluster files should be read and written by the `postgres` user, while other users shouldn't have any permissions (0700 mode). If this is not the case, PostgreSQL won't start correctly.

To use continuous archiving, we need a base version of our cluster. This base version is the first checkpoint. From this checkpoint, logs are preserved, and we can restore previous states by restoring the first checkpoint, specifying a date, and letting PostgreSQL replay the logged transactions until the specified date. To enable WAL archiving, we have to set some system variables using a superuser role as follows:

1. Set the `wal_level` variable to `archive` with the expression `ALTER SYSTEM SET wal_level = 'archive';`.

2. Set the `archive_mode` variable to `on` with the expression `ALTER SYSTEM SET archive_mode = 'on';`.

3. Create a place for your archives. Remember the absolute path to that place. I will use the `/home/debian/postgres_archive` path.

4. Set the `archive_command` variable to the system call that PostgreSQL should archive WALs with. On Linux and macOS systems, it can be `ALTER SYSTEM SET archive_command = 'test ! -f /home/debian/postgres_archive/%f && cp %p /home/debian/postgres_archive/%f';`, while, on Windows, it should be something like `ALTER SYSTEM SET archive_command = 'copy "%p" "C:\\postgres_archive\\%f"';`. In the call, `%f` denotes the WAL file's name, while `%p` denotes its absolute path with its name.

5. Restart the server.

> Telling PostgreSQL what to do in the archiving process might seem tedious, but it gives an amazing amount of flexibility. We can encrypt or compress the WALs, send them through SSH, or do virtually any valid operations on them.

Next, we have to set up the first checkpoint, and create a physical copy of this base version. We can put this backup wherever we like, although it should be placed somewhere along the WAL files.

1. Start creating the first checkpoint with the query `SELECT pg_start_backup('backup', true);`. By specifying `true`, we ask PostgreSQL to create the checkpoint as soon as possible. Without it, creating the checkpoint takes up about 2.5 minutes with the default settings. Wait for the query to finish.

2. Copy out everything from the cluster to the backup folder. You can use any tool for this, although you must make sure that file permissions remain the same. On Linux and macOS, `tar` is a great tool for this. With my paths, the command looks like the following:

```
tar -czvf /home/debian/postgres_archive/basebackup.tar.gz
  /database/postgres.
```

3. Stop the backup mode with the query `SELECT pg_stop_backup();`.

 There is a CLI tool called `pg_basebackup`, which can automatically create the first checkpoint and its backup. However, it needs PostgreSQL to be configured in a way that replication connections are accepted. For further reference, you can read the official manual at `https://www.postgresql.o rg/docs/9.4/static/app-pgbasebackup.html`. You can also read a thorough guide on configuring a hot standby server at `https://cloud.go ogle.com/solutions/setup-postgres-hot-standby`.

Let's say the worst has happened, and our database is corrupted. In that case, our first task is to find out the last time our database was stable. We can guess, but in this case, guessing is a bad practice. We should look through our logs to see when our database went off. We don't want to have to rollback more transactions than necessary, as this can have a significant impact on a production server. When we have a date, we can start recovering with PostgreSQL's **PITR (Point-in-Time Recovery)** technique:

1. Shut down the PostgreSQL server.
2. Make a backup copy from the corrupted cluster's `pg_xlog` folder, as it might contain WAL files which haven't been archived yet. It is a good practice to make a copy of the corrupted cluster for later analysis if you have the required free disk space.
3. Delete the cluster, and replace it with the base backup's content.
4. The base backup's `pg_xlog` folder's content is now obsolete, as those changes were already incorporated in the backup database. Replace its content with the corrupted cluster's logs. Watch out for keeping the correct permissions!

5. Create a file named `recovery.conf` in the cluster. The file's content must contain the inverse of the archiving command saved to the `restore_command` variable. It should also contain the date until the recovery should proceed to be saved to the `recovery_target_time` variable:

```
restore_command = 'cp /home/debian/postgres_archive/%f %p'
recovery_target_time = '2017-02-21 13:00:00 GMT'
```

 You can read more about the valid date formats PostgreSQL accepts at `http s://www.postgresql.org/docs/9.4/static/datatype-datetime.html`.

6. Start the server. When PostgreSQL is done with the recovery, it will rename `recovery.conf` to `recovery.done`.

Summary

In this chapter, we discussed PostgreSQL's structure, its objects, and how PostGIS sits on the RDBMS. We also learned some of the architectural specialities of PostGIS. We came closer to fully understanding RDBMSs, what we should look out for when we use them, and how we can effectively create queries in them. Although we used pgAdmin, we also learned some useful expressions, which can be used directly in PostgreSQL's CLI. It will come in handy when you have to configure a PostgreSQL or PostGIS instance on a remote server only accessible through SSH.

In the next chapter, we will dive into geospatial analysis, and see how we can produce meaningful results from our raw data. We will set up a scenario where we are real estate agents serving a customer with very specific needs. To find out the best spots matching the given criteria, we will use various geoalgorithms via geoprocessing tools in QGIS.

8
Spatial Analysis in QGIS

In the previous chapter, we learned how spatial RDBMSs work on the example of PostGIS. We went through the typical data types, tables, views, functions, and other objects, while also discussing how we can construct a great spatial database, which can effectively enhance our work. Now that we can ease our job by using spatial databases, we will move on and learn about spatial analysis. To have a full-blown example on which we can work, let's assume we are very enthusiastic real estate agents. We are happy to go that extra mile to please our customers who have very specific conditions on their dream houses. The price and size of the house does not matter in this case, although its situation must meet certain criteria.

In this chapter, we will cover the following topics:

- Vector analysis
- Building models in QGIS
- Digital elevation models (DEM)
- Using GRASS from QGIS

Preparing the workspace

First of all, we should choose a small area to work with. A populated town or city is an obvious choice for this task. The first obstacle is that there aren't any freely available settlement polygon data in the formats we are used to. To tackle this, we can download the required data directly from OpenStreetMap. OSM offers a read-only web database for accessing its data, which is available through the Overpass API. From Overpass, we can request data through regular web requests, and the server sends the matching features as a response (*Appendix 1.5*). In QGIS, we can install a plugin written directly for this-- `QuickOSM`.

Therefore, our first task is to install the `QuickOSM` plugin as follows:

1. Open the plugin manager via **Plugins | Manage and Install Plugins**.
2. Type `QuickOSM` in the search field.
3. Click on **Install plugin**.

In `QuickOSM`, we can build Overpass requests in an interactive way. In Overpass, we can query features in a predefined area (for example, the extent of a layer), or use a name which can be geocoded to an area by OSM (for example, the name of an administrative boundary). If we open up the plugin, we can see that Overpass accepts key-value pairs according to OSM's tag specification. If we would like to request administrative boundaries, we have to use the `admin_level` tag as **Key**, and provide the appropriate level as a **Value**. We only need to find out the level for our country. Luckily, OSM maintains a detailed wiki about the tags it uses from where we can reach the `admin_level` related part at `https://wiki.opens treetmap.org/wiki/Tag:boundary%3Dadministrative#admin_level`:

Greece (proposed)	N/A	Εθνικά Σύνορα	N/A	Όρια Αποκεντρωμένων Διοικήσεων	Όρια Περιφερειών (NUTS 2)	Όρια Περιφερειακών ενοτήτων (NUTS 3)	Όρια Δήμων (LAU 1)	Όρια Δημοτικών ενοτήτων (LAU 2)
Guatemala (proposed --Esteban Ortiz 21 June 2009 (UTC))	N/A	National border (Frontera nacional)	N/A	State border (Departamentos)	N/A	Municipal border (Municipios)	N/A	City and town border (Zonas)
Haiti (Republic of) (proposed --Vaandre 16:19, 15 January 2010 (UTC))	N/A	National border	N/A	State border (Départements) (Layer 1)	Districts (Arrondissements)	Search-And-Rescue sectors (**temporary**)	N/A	City and town border (communes) (Layer 2)
Honduras (proposed --Antonio Locandro 11 March 2014)	N/A	Fronteras (National Border)	N/A	Departamentos (State Border)	N/A	Municipios (Municipal Border)	N/A	Aldeas (Admin level border which encompass several towns/cities)
Hungary (proposed/used)	N/A	Országhatár (national border)	N/A	Országrészek (Groups of Regions, NUTS 1)	Régiók (Regions, NUTS 2)	Megyék / főváros (Counties / capital city, NUTS 3)	Kistérségek, járások (LAU 1)	Települések (LAU 2)
Iceland	N/A	Landhelgi - Territorial waters	N/A	N/A	Regions	Sveitarfélög - Municipalities	N/A	N/A
India (see also administrative divisions)	N/A	National Border	Division	State	District	Subdistrict (Tehsil / Mandal / Taluk)	Metropolitan Area	Municipal Corporation / Municipality / City Council
Indonesia	N/A	National Border	N/A	Province	City / Regency (Kotamadya / Kabupaten)	Subdistrict (Kecamatan)	Village (Kelurahan / Desa)	Hamlet (Dusun)
Iraq (proposed by johanemilsson, modified Łukasz (talk) 15:58, 24 August 2016 (UTC))	N/A	National Borders	Iraqi Kurdistan (Arbil, Duhok, Sulaymaniyah) region	governorates (muḥāfaẓah)		districts (qadha)	subdistricts (nahya)	boroughs (hay) in urban and counties (mukataa) in rural areas

Now we can import our preferred city with the following steps:

1. Open up the `QuickOSM` plugin.
2. Provide `admin_level` as **Key**.
3. Find out the level that contains settlement data for your country from the aforementioned link, and provide that level as **Value**.

4. Check the radio box next to **In**, and provide the settlement's name.

5. Expand the **Advanced** menu.

6. Keep only the **Relation** and the **Multipolygons** boxes checked. This way, we will only request polygons from OSM.

7. Click on **Run query**:

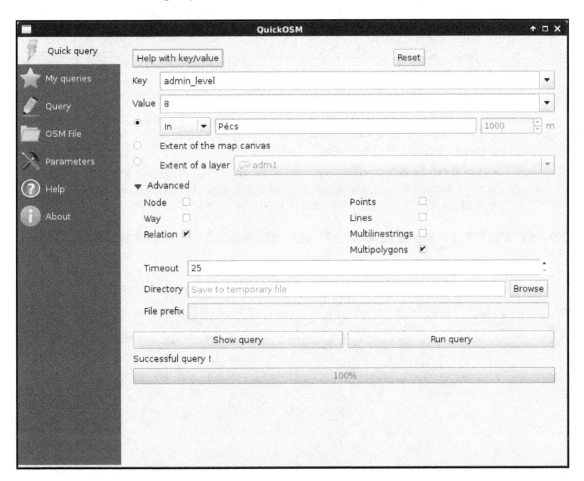

If running the query yields an error, restart QGIS, click on **Reset** in `QuickOSM`, and fill out the form again. If you get the correct result, saving it in your working folder is recommended. If your town's name is not unique (for example, Los Angeles), you might have to filter the results.

As we have a smaller region for our analysis, let's generate some data. We should create points representing houses for sale with street addresses, sizes, and prices. There are various ways to achieve this. As we already did some geoprocessing, please take a little break to think about a sequence of logical steps that you would take to create these random data.

In my opinion, the easiest way is to generate the points right on the streets. We can do that as follows:

1. Load the `roads` layer.

2. Apply a filter on the layer which discards streets without the `name` attribute. Use the filter expression `"name" IS NOT NULL`.

3. Clip the `roads` layer to the town boundary. Use **QGIS geoalgorithms | Vector overlay tools | Clip** for this task. Save the clipped layer as a memory layer by specifying **memory:** as output.

4. Generate random points on the clipped road layer with **QGIS geoalgorithms | Vector creation tools | Random points along line**. Create at least 1,000 points, so we will always have some results. Save the points as a memory layer:

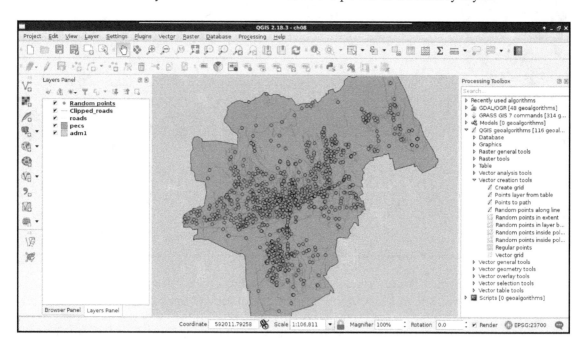

The next step is to fill our random points with data. First of all, we need to join the attributes of the roads layer to have the street names stored with our imaginary houses. The only problem is that QGIS's spatial join uses the problematic **Select by location** tool's algorithm. To see what I'm talking about, let's select points with it (**QGIS geoalgorithms** | **Vector selection tools**) using the **intersects** spatial predicate and our clipped lines layer as the intersection layer. For me, it did not select any points. One of the problems might be a precision mismatch between lines and points interpolated on them (*Appendix 1.6*); however, if we define a small **Precision** value, still only a portion of our points get selected. For me, a precision of 170 meters resulted in a full selection, where QGIS has no means to guarantee the correct street attributes get joined to the points.

What we already know about spatial queries in QGIS--checking point-polygon and line-polygon topological relationships--work well. Checking point-line relationships, on the other hand, does not. Therefore, the easiest workaround is to transform our points to polygons, join the street attributes to them, then join the polygon attributes back to the points. For this task, we can use the basic geoprocessing tool--buffer. It creates buffer zones around input features; therefore, if we supply a point layer to the tool, it creates regular polygons around them. The more segments a buffered point has, the more those polygons will resemble circles. There are two kind of buffer tools in QGIS--fixed buffer and variable buffer. The variable buffer tool creates buffer zones based on a numeric attribute field, while the fixed buffer tool creates buffer zones with a constant distance. Let's create those joins using the following steps:

1. Select the **Fixed distance buffer** tool from **QGIS geoalgorithms** | **Vector geometry tools**.
2. Supply the random points as the input layer, and a small value as **Distance**. A value of 0.1 meters should be fine. Save the result as a memory layer.
3. Use the **Join attributes by location** tool from **QGIS geoalgorithms** | **Vector general tools** for the first join. The target layer should be the buffered points layer, while the join layer should be the clipped roads layer. The spatial predicate is **intersects**. Save the result as a memory layer.
4. In this step, we can either use the same tool to join the attributes back to the `random points` layer, or do ourselves a favor and use a simple join. As both the tools preserved the existing attributes, we have the original IDs of our random points on the joined buffered points layer. That is, we can specify a regular join in the original `random points` layer's **Joins** tab in its **Properties** menu. When we add a new join, we should specify the joined buffered points layer as the `Join layer`, and the `id` field as the join and target fields. Finally, we should restrict the joined columns to the `name` column, as we do not need any other data from the `roads` layer.

5. Open the **Field calculator** for the `random points` layer. Create a new integer field named `size`, and use QGIS's `rand` function to fill it with random integers between two limits. If you would like to create sizes between 50 and 500, for example, you have to provide the expression `rand(50, 500)`.

6. Create another integer field named `price`, and fill it with random numbers in a range of your liking. As I'm using my currency (HUF), I provided the expression `rand(8000000, 50000000)`. Watch out for the **Output field length** value, as the upper bound should fit in the provided value.

7. Exit the edit session, and save the edits made to the points layer. Save the layer in your working folder, and remove all other intermediate data (including the clipped roads):

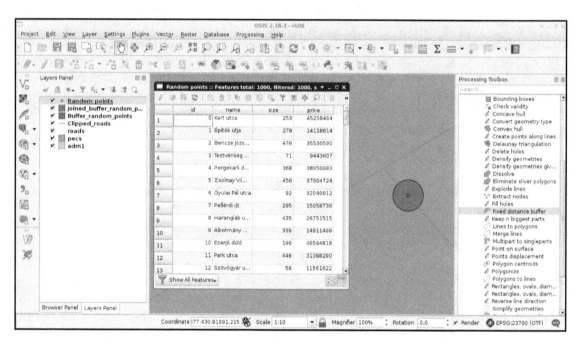

Laying down the rules

Two customers walk into our agency. They state their preferences, share some contact information, and leave. When we summarize their preferences, we are amused--the vague terms are totally identical. Both the customers stated the following criteria:

- It shouldn't be too noisy
- There should be at least a restaurant and a bar nearby

- There should be some markets nearby
- There should be a park nearby, preferably with a playground

Additionally, both of them had a single, more specific request.

- **Customer 1**: I would like to place some solar panels on the rooftop, and use them efficiently.
- **Customer 2**: I don't have a car, and prefer to walk rather than use public transport. However, I like to travel, therefore, the train station and the bus station should be at a walking distance.

Our first task in such a case is to interpret the criteria, and translate them into the language of GIS. We have to create some exact steps from these preferences. Of course, having such vague criteria always leaves space to some subjectivity; therefore, the following interpretation accords to my experience:

- **It shouldn't be too noisy**: The results should be more than 500 meters away from industrial areas, and more than 200 meters away from busy roads (like motorways and highways) in terms of linear distance.
- **There should be some amenities, shops, and a park with a playground nearby**: The results should be less than 500 meters away from at least one park with a playground, bar, restaurant, and from two markets in terms of linear distance.
- **Efficient solar panels**: The aspect of the results' area should face South in the northern hemisphere, and face North in the southern hemisphere.
- **Stations at walking distance**: The cost of walking to those stations from the results must not exceed 15 minutes. If this criterion cannot be met (that is, the train and bus stations are too far away), or there are no results (that is, the intersection of the valid costs is too small to contain any results), we will use the union of the valid costs.

Implementing airport noise into the model is also possible, although our current data is not sufficient for this. To estimate airport noise pollution, we would need at least the paths of taking off and landing airplanes, and buffer them.

Vector analysis

We can get a partial result by only considering our vector layers, and running some vector analysis tools on them. There are different methods for different type of analysis; however, we can group the most frequently used methods into the following four groups:

- **Overlay analysis**: Analyzing features according to their spatial relationships to other features. Common use cases are spatial queries and spatial joins.
- **Proximity analysis**: Analyzing the relationship of features based on some distances. The heart of this type in a traditional desktop GIS software is the buffer tool, while the rest of the work is basically overlay analysis.
- **Neighborhood analysis**: Analyzing (more often, statistically) neighbouring features of some input features. When we need to find the closest features to some input features, it is called a **k-NN** (**k nearest neighbor**) query.
- **Network analysis**: Analyzing a topological network, or some features on it. The most typical use case is to find the shortest path between two points on a road network.

Different GIS softwares are good at different analysis types. Although QGIS is a universal GIS, it offers a good coverage only in the most basic analysis types--overlay analysis and proximity analysis. PostGIS is exceptionally good in neighborhood analysis. As GRASS GIS forces the topological model on vector layers, it has the best capabilities in network analysis.

If you would like to learn about network analysis, you can look up examples using pgRouting with PostGIS, or using network tools in GRASS GIS.

Proximity analysis

To fulfill most of our criteria, simple proximity analysis is enough. However, there are two types of criteria. Some of them state that our results have to exceed a distance, while the rest of them require the results to equal a certain distance.

For the sake of simplicity, let's separate those requirements, and solve them one by one. The easiest way to do a proximity analysis with fixed distances is to buffer the features we want to compare our houses with, and use a spatial query to get matching results. We can delimit houses far enough from noise-polluted areas by the following (from now on, saving intermediate results as memory layers won't be emphasized):

1. Open the `roads` and `landuse` layers.
2. Apply a filter on the `roads` layer to show only relevant roads. The correct expression is `"fclass" LIKE 'motorway%' OR "fclass" LIKE 'primary%'`.
3. Apply a filter on the `landuse` layer to show only industrial areas. Such an expression can be `"fclass" = 'industrial' OR "fclass" = 'quarry'`. If you have other types, which can be a source of noise pollution, don't hesitate to include them.
4. Clip the layers to the town boundaries if they are too large (optional step).
5. Buffer the roads with the **Fixed distance buffer**, tool and a buffer distance of 200 meters. If you have a projection in feet, the correct value is 656. If your CRS is using miles as the unit, it is 0.12.
6. Buffer the `landuse` layer with a value of 500 in meters, or the equivalent value in other units.
7. Get the union of the two buffered layers with **QGIS geoalgorithms | Vector overlay tools | Union**. The order does not matter. We need a union of the two layers, as neither of the buffer areas are suitable for us.
8. Save the features outside of the result with the **Extract by location** tool. We should select from the house layer, use the noisy places layer as the intersection layer, and select **disjoint** as the spatial predicate.

9. Remove the intermediate layers, and save the filtered houses if you plan to follow this chapter in multiple sessions:

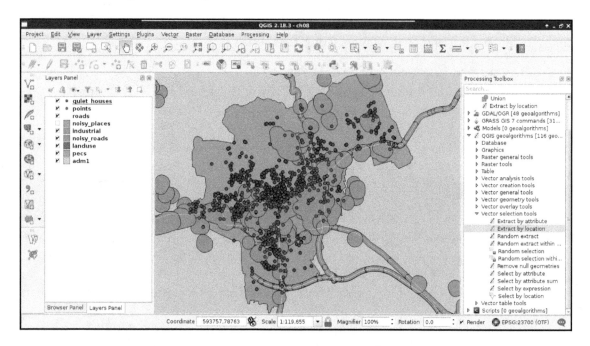

The second part is to delimit the areas that the potential homes should reside in. The first criterion is parks with playgrounds. Let's prepare our data as follows:

1. Apply a new filter on the `landuse` layer. The filter should be `"fclass"` = `'park'`.

2. Open the `POI` layer, and apply a filter which only shows playgrounds. The correct expression is `"fclass"` = `'playground'`.

   Now we could select or extract parks which have a playground in them. Before doing this, let's think it through again. Is this the correct method? What if there are some parks with playgrounds just outside of them?

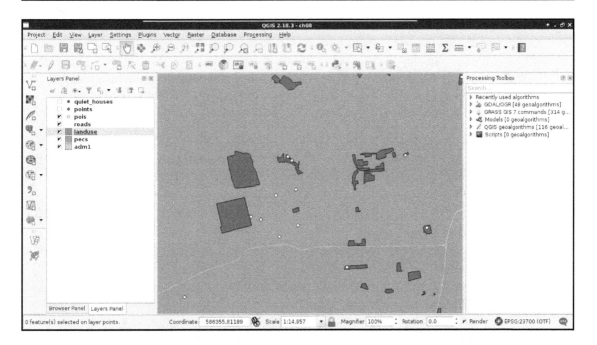

3. In order to apply a more permissive extraction, use the **Select by location** tool, and select features from the `landuse` layer with respect to the `POI` layer. Select **intersects** as a spatial predicate, and include a **Precision** value somewhere between 100 and 200 meters.

 Selecting or extracting features with a precision value is very convenient, although it cannot replace buffering in QGIS when precise results are required (*Appendix 1.7*).

4. Buffer the `landuse` layer with the selected parks by 500 meters. Check the **Dissolve result** box. Remember that QGIS's geoalgorithms respect filters? Well, most of them also respect selections. Therefore, the result only contains buffered versions of the selected parks only.

5. Remove the selection with the **Deselect Features from All Layers** button in the main toolbar.

Wondering if you should dissolve the buffered features automatically? Well, if you need to keep the attributes associated to individual features for a later analysis, you shouldn't. If not, it depends. Dissolving features causes some overhead, although you get a much cleaner result. If you are going to make some overlay analysis on the buffered layers (like intersecting them), those operations will run faster on dissolved buffer zones.

6. Apply a new filter on the `POI` layer considering only bars. I'll let you construct the correct expression this time. Think it through well; pubs, for example, can be considered bars in this case. Can cafes be considered bars as well? I'm going to leave this to you.

7. Buffer the filtered `POI` layer with 500 meters, and dissolve the buffered features automatically.

8. Apply a new filter on the POI layer which shows only restaurants. Such a filter expression is `"fclass" = 'restaurant'`.

9. Buffer the filtered `POI` layer with 500 meters, and dissolve the buffered features automatically.

10. Now that we have three buffered constraint layers, we only need areas which fulfill all the criteria. Therefore, we need the intersection of the three buffered layers. We can calculate the intersection of two layers with **QGIS geoalgorithms | Vector overlay tools | Intersection**. Let's create the intersection layer of the buffered bars and the buffered parks layers.

11. As the **Intersection** tool only accepts two layers, we have to intersect the third layer (buffered restaurants) with the result of the previous step. Let's do that to get the final constraint layer:

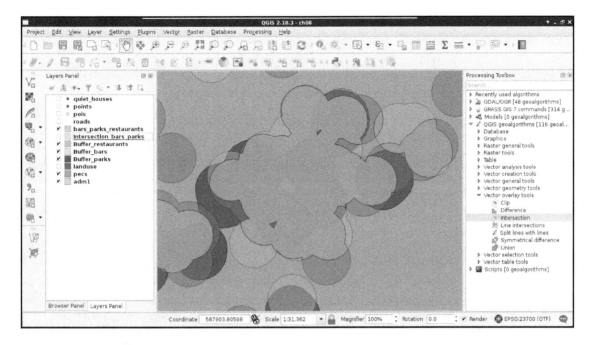

12. As you can see in the preceding screenshot, the final intersection layer only contains areas which are present in every buffer layer. Now extract every house from the quiet houses layer, which intersects the constraint layer. Use the **Extract by location** tool.

13. Remove every intermediate layer. Save the constrained quiet houses layer on the disk if you are planning to follow the rest of the chapter in another session.

Understanding the overlay tools

You must be wondering about a lot of things now. First of all, what's the difference between clipping and intersecting two layers? Not much, to be honest. Clipping is a special type of intersecting, where the second layer must be a polygon layer, and the attributes of the second layer don't get included in the input layer. Intersecting, on the other hand, can be used on any vector type. Furthermore, the attributes of the second layer are automatically joined to the input layer. To sum it up, clipping is a special type of intersection, where the algorithm knows that we only want to restrict a layer's geometries to some bounds, and nothing more.

As we stated before, if we use buffers for proximity analysis, half of it will be overlay analysis (checking if the input features reside in the proximity). One of the main use case of overlay analysis in proximity analysis is to create the right bounds for the upcoming selection or extraction. This work is basically set operations in two or more dimensions. Let's say we have a polygon **A** in a layer and a polygon **B** in another one.

- **Intersection (A ∩ B)**: We search for every point in our data frame which can be found in both **A** and **B**
- **Union (A ∪ B)**: We search for every point which can be found in either **A** or **B**
- **Difference (A - B)**: We search for every point which can be found in **A**, but cannot be found in **B**
- **Symmetrical difference (A Δ B or (A - B) ∪ (B - A))**: We search for every point which can be found in one of the inputs, but cannot be found in the other

Understanding why we used intersections in the second task is easy. We needed the intersections of the buffer zones, as each zone contained a specific criteria from which every one should be fulfilled by the results. However, why did we use union in the first task? The answer lies in negating, as it turns the logic upside down. Think of it like this--we need every house in the intersection of areas outside the two buffer layers. According to De Morgan's law (the intersection of two sets' complement is the complement of their union), we need every house in the area outside of the union of the two buffer layers.

Towards some neighborhood analysis

The last criterion that both customers need is that the number of markets in the vicinity of the house should be at least two. This is another type of proximity analysis, as we do not need a binary answer (it is in the proximity of the other feature, or it isn't), we need to count the number of features. Count is one of the most basic statistical indicators in GIS. In my opinion, this step is somewhere between proximity analysis and neighborhood analysis (where we do not even care about proximity, just the distance).

We have several ways to achieve this step. We can buffer the markets without dissolving them, and execute a spatial query with statistics. On the other hand, that is not a clear way to get only the number of markets, as the result would contain some other statistics we do not care about. To get only counts, we can do this in reverse:

1. Apply a new filter on the `POI` layer to only show markets. There are a lot of shop types in OSM, therefore, the expression can vary from place to place. Such an expression is `"fclass" = 'supermarket' OR "fclass" = 'convenience' OR "fclass" = 'mall' OR "fclass" = 'general'`. You can read more about shop tags at `https://wiki.openstreetmap.org/wiki/Key:shop`.

2. Buffer the extracted houses--on which we've already applied most of the constraints in the previous steps--with 500 meters. Do not dissolve the result.

3. Count the number of markets in the individual buffer zones with **QGIS geoalgorithms | Vector analysis tools | Count points in polygon**. The count field name can be anything; I will name it `count`.

4. Join the new polygon layer containing the number of markets back to the house layer with a regular join (**Properties | Joins**). Both the join and target fields should be `id`, while the joined columns should be restricted to the `count` column.

5. Select and save houses which have at least two markets in their vicinity. Use the expression `count >= 2`, and either select them using **Select features using an expression**, then saving them with **Save As**, or use the tool **Extract by attribute**.

6. Remove intermediary layers (including quiet houses and constrained houses from the previous steps):

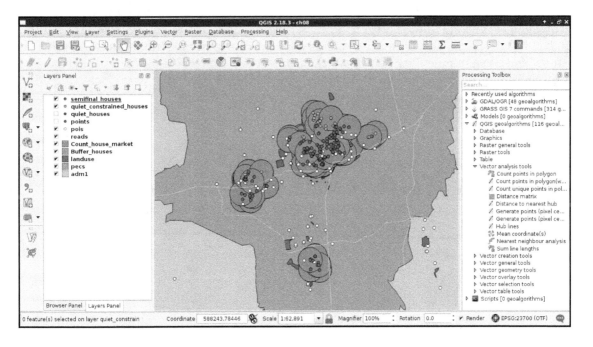

Building your models

Congratulations on your first analysis! It was quite an adventure, right? What we've done is more than mere spatial analysis. We conceptualized a model, and made an analysis according to that. Our model stated that the vicinity of the requested amenities and features can be translated to 500 meters. Quiet places are places which are more than 200 meters away from busy roads, and more than 500 meters away from industrial places. Are these numbers exact? Of course not. They are approximations of real-world phenomena, and therefore, models.

What happens if one of the customers says that our analysis is faulty? Some of the results are too close to noisy places, others are too far from markets. We can try some other distances to make our model satisfy the customer better, although we would need to run the entire analysis every time. Luckily, in modern desktop GIS software like QGIS, there is a graphical modeler to create, save, and modify a step-by-step analysis by connecting algorithms to each other. It is like a block-based programming language for analysts. We can link existing algorithms (even models) together to create a graphical process model, that QGIS then interprets and executes.

We can access QGIS's graphical modeler from the menu bar via **Processing** | **Graphical Modeler**. First of all, we need to name our model, and categorize it in a group. I used the name House search and the group Vector. If we save a new model, we have to specify a file name, which can be anything as long as we don't change the default directory QGIS offers, and use a unique file name. If we close the model, we can see it under the category we specified. We can edit our existing models by right-clicking on them, and selecting **Edit model**:

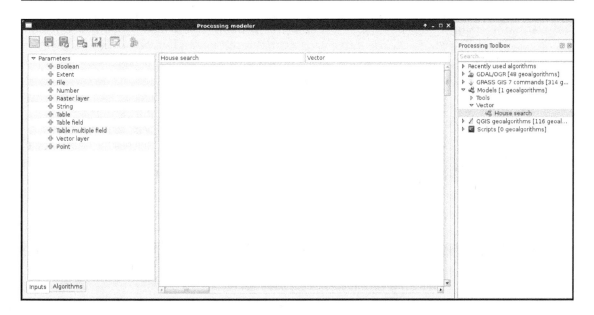

The graphical modeler has a lot of capabilities from which we will only use the most necessary ones to create our model. The left panel shows the inputs and algorithms we can use. We can simply drag and drop the needed blocks to the right panel, which is the canvas of our model. As the first step, let's create the quiet homes part. For this, we need three input vector layers--a point layer for the `Houses`, a line layer for the `Roads`, and a polygon layer for `Land use`. When we drop a **Vector layer** input to the canvas, we can specify the name and the type of the input layer:

Now, if we save our model, and run it with the **Run model** button or by opening it from the processing toolbox, we can see our three constrained input vector layers just like in any other QGIS algorithm. Now we need to drag and drop some algorithms from the **Algorithms** tab of the left panel, which will use our input layers:

1. Drag in the first algorithm--**QGIS geoalgorithms** | **Vector selection tools** | **Select by expression**. Select the `Roads` layer as an input layer, and provide the expression `"fclass" LIKE 'motorway%' OR "fclass" LIKE 'primary%'`. Give it the description `Select busy roads`.

2. Drag in a **Fixed distance buffer** algorithm. Select the output of the previous tool as an input, and define a buffer zone of 200 meters. You can also dissolve the result. Give it a description, something like `Busy roads buffer`:

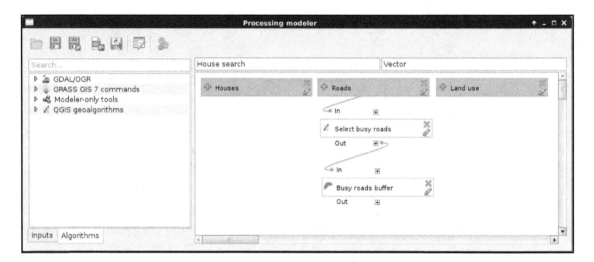

If you need to change a parameter or an algorithm, you can click on the pencil icon in the lower-right corner of its box. If QGIS does not respond, you can right-click on the box, and select **Edit**. To remove an item, you can right-click on it, and select **Remove**. You can only remove items from the end of the processing chain.

As we can see, we have access to some extra features besides the regular parameters that QGIS offers in the graphical modeler. These include the following:

- **Description**: We can describe an algorithm, as the graphical modeler can hold multiple instances of the same tool. This way, we can distinguish between them when we build the rest of our model. Always add a unique description.
- **Parameters**: These are the regular parameters that QGIS requires.

- **Output**: Some of the algorithms can produce an output. If we give it a name, QGIS treats it as a result, and offers us to save it somewhere. If not, QGIS knows that it is just an intermediary step producing temporal data.
- **Parent algorithms**: We can affect the order of execution by setting additional parent algorithms of a geoalgorithm.

> Although the **Select by expression** algorithm operates only in place, you can export the selected features with the **QGIS geoalgorithms** | **Vector general tools** | **Save selected features** tool.

Let's finish modeling the first step of our analysis with the following steps:

1. Add another **Select by expression** tool. The input should be the Land use layer this time, while the expression is `"fclass" = 'industrial' OR "fclass" = 'quarry'`.
2. Add another **Fixed distance buffer** tool. The input should be the selected land use layer, and the buffer distance should be 500 meters. You can dissolve the result.
3. Add a **Union** tool. The two inputs should be the two buffered layers. Specify an output name, as we can test our model that way.
4. Save the model, and run it:

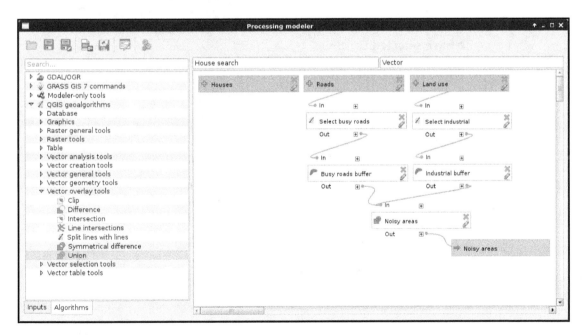

Now we can see something, which highly resembles the geoalgorithms we are used to in QGIS. It requires three inputs, and gives one output. Let's remove the filters from the required layers, specify them as input parameters, and run the query.

Do not specify **memory:** as output. Models cannot produce memory layers in QGIS. If you do not want to save the result, leave the output field blank, and let QGIS create a temporary layer.

By running the model, we can notice a few things. First of all, the result is similar to the unified buffer zones we created step by step. However, we can do the whole workflow by simply pressing a single button. However, the model is seemingly quite slow. More precisely, dissolving the buffers slows down the whole process. We can do these few things about that:

- We can disable dissolving, which will make buffering faster, but union slower.
- We can build a geometry index on the inputs of the buffers with **QGIS geoalgorithms | Vector general tool | Create spatial index**.
- We can save the selected features or use the **Extract** tools instead of the **Select** tools. QGIS models and PostGIS selections are not the greatest duo, especially when a spatial query follows, because they decrease performance.

Note that we do not have an **Extract by expression** tool (*Appendix 1.8*) in QGIS. If an arbitrary SQL expression is needed, the only workaround in the graphical modeler is to use **Select by expression**, and export the result with **Save selected features**.

For now, let's just finish the current part of the analysis:

1. Edit the model.
2. Remove the output produced by the **Union** tool. You just have to remove the text from the **Union<OutputVector>** field.
3. Add an **Extract by location** algorithm. We should select from the Houses layer, specify the unified buffers as the intersection layer, and **intersects** as the spatial predicate.
4. Specify an output to the **Extract by location** algorithm.
5. Rename the model to something like Quiet houses, and the group to Real estate analysis.
6. Save the model.

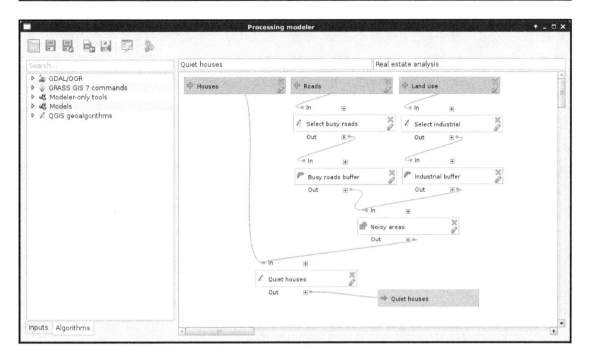

Now we have the Quiet houses produced by our model. The next part is to constrain those houses with the preferences of our customers. To keep our final model clean, we are going to separate different tasks. Let's create another model with a name like Constrained houses, and with the previous model's group name:

1. Add three input vector layers--one for the Houses, one for the Land use, and one for the POI layers.

2. Select the parks from the Land use layer, and the playgrounds from the POI layer. Save the selected features. As both the selections only take a single key and value, you can use a single **Extract by attribute** tool instead ("fclass" = 'park' from the Land use layer and "fclass" = 'playground' from the POI layer).

3. Select parks in the vicinity of playgrounds, and buffer the results.

4. Select bars from the POI layer, and buffer the results.

5. Select restaurants from the POI layer, and buffer the results.

6. Intersect the buffered layers. First take two of them as the input of an **Intersection**, and then take the result with the third buffered layer as inputs of another **Intersection**.

7. Extract houses located in the final result. Name the output of this final step:

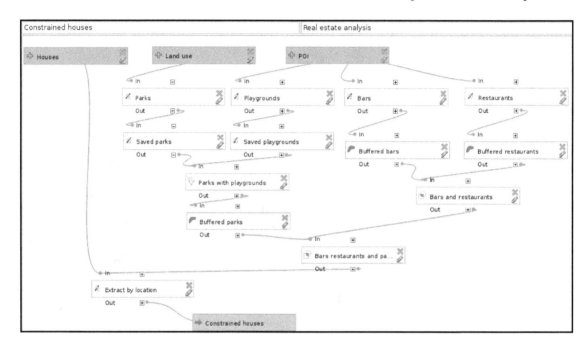

Our final model should contain every input, our other models, and the rest of the analysis:

1. Create a new model with a name something like `House search` or `House analysis` in the same group.
2. Create four inputs for the house, road, land use, and `POI` layers.
3. Add the `Quiet houses` model, and specify the inputs.
4. Add the `Constrained houses` model specifying the output of the previous model as the `House` layer input, and the rest of the input layers as the other inputs.
5. Buffer the result of the previous model.
6. Select the markets from the `POI` layer, and save the selection.
7. Use **Count points in polygon** to count the number of markets in the houses' buffer zones.
8. Join the output of the `Constrained houses` model with the result of the previous tool by using **QGIS geoalgorithms | Vector general tools | Join attribute table**. Both the table fields should be `id`.
9. Extract the valid features by using the expression `count >= 2`. We can use the **Extract by attribute** tool for this:

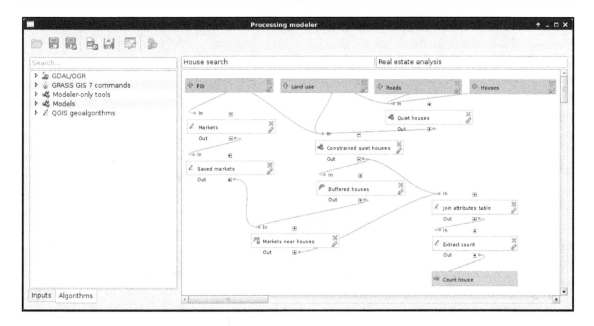

Let's test our model by running it and examining the results. If something really weird did not occur, we've got bad results. Not just slightly **I must have done something wrong in my step-by-step workflow** bad results, but really bad ones. Why did something like this happen? In a nutshell, QGIS does not have a concept about correct order. It interprets our model, and orders the algorithms based on inputs. Every algorithm has so many dependencies as inputs, which must be executed before them. Other algorithms are executed in an arbitrary order, which is good, as QGIS 3 will be able to run processing algorithms in parallel.

I'm sure you already found out the solution for this ordering problem--we must make sure that order does not matter. This can be achieved by chaining algorithms in a way that our steps do not rely on the state of the data. If we think it through, in our model, data has state only in a few cases; that is, when we use selections. We have the following two ways to resolve this problem:

- We can discard selections, and work with extraction algorithms where we can. Where we cannot, we can build an `Extract by expression` model (*Appendix 1.8*), and use that instead.
- We can force correct ordering by defining additional parent algorithms for some of our steps.

Let's stick with the second option now. If we think about the possible orders of execution in our models, we can conclude that the `Quiet houses` model is safe. No matter in which order QGIS executes it, we will always get the same results. On the other hand, there are several incorrect paths in the `Constrained houses` model, and an additional one in our final model. In the `Constrained houses` model, we select three times from the same `POI` layer. If the second selection (bars) occurs after the playgrounds are selected, but before they are saved, we get incorrect results. Let's correct it by defining the `Saved playgrounds` step as a prerequisite to the `Bars` step:

1. Edit the `Constrained houses` model.
2. Edit the `Bars` step with the pencil icon, or by right-clicking on it and selecting **Edit**.
3. Click on the chooser (...) button besides the **Parent algorithms** field.
4. Select the `Buffered Parks` algorithm. We do not have the unique names we gave to our steps in this dialog, although when we select an algorithm, QGIS connects the two steps together with a grey line. Check the result, and if the wrong algorithm got connected, try again.

Using the same procedure, we should make the `Buffered bars` step a prerequisite to the `Restaurants` step, as that is the second place where an error can occur. When done, let's inspect our final model. There we select the markets, and save them to another layer. However, what happens if our second model runs after the selection, but before the extraction? The correct selection is gone, and a wrong selection gets saved. To deal with this possibility, we have to make our `Constrained quiet houses` step a prerequisite to our `Markets` step. If we run our model again, the results should be the same as the ones gained from our step-by-step approach.

Don't worry if the two results are not entirely the same. You can easily get slightly different results from a manual workflow just by mistyping some required parameters or fiddling with the optional ones. An appropriately validated model is always more reliable than a long manual workflow.

Using digital elevation models

The third preference of both the customers are quite special ones, which cannot be incorporated into our model with mere vector analysis tools. One of the requirements is that solar panels can be used on the house efficiently. Of course, there are numerous factors to consider, like the exposure of the roof.

However, the simplest feature such a house should have is the aspect of the land it is built on. If we are in the northern hemisphere, the terrain with northern aspect is mostly in shadow. In the southern hemisphere, the land with southern aspect is more shadowy. In reality, there are a lot more factors contributing to the solar energy potential (like latitude, climate, cast shadow, and so on), but for the sake of simplicity, let's just assume for now that the only factor is the aspect of the surface.

For solving surface-related problems, GIS has its most characteristic data type--**DEM (Digital Elevation Model)**. DEMs are representations of a planet's surface (most often Earth). They can be in raster or vector (mesh) format, and can be visualized in 2D, or as a rubber sheet in 3D, which is not 2D, but neither is it true 3D (they are often called 2.5D because of this property). What makes DEMs a very valuable and useful data type is their wide variety of use cases. A lot of terrain-related information can be derived solely from the surface data. The two most basic derivatives are slope and aspect, where slope shows the steepness of the surface, while aspect shows its exposure.

DEMs in a regular grid format (raster) are usually the results of processing raw elevation data in vector format. These raw elevation data can be acquired in multiple ways, such as point clouds from RADAR or LiDAR measurements, digitized elevation contours, or individual GPS measurements. If the resulting vector data is dense enough, a regular grid can be constructed on them, and the individual data points in a cell can be averaged. If not, then spatial interpolation is the usual way of creating DEMs. There are a lot of spatial interpolation techniques, although there is a common concept in them--they take a number of points as input, and create a regular grid by interpolating additional points between existing ones as output (*Appendix 1.9*). Of course, these irregular elevation points can also form a **Triangulated Irregular Network** (**TIN**) if their Delaunay triangulation is calculated.

There is a lot of additional knowledge about digital elevation models, which is out of the scope of this book. If you wish to learn more about DEMs, you can start with Wikipedia's article at `https://en.wikipedia.org/wiki/Digital_elevation_model`, then continue with more serious writings like *Terrain analysis--principles and applications* by John P. Wilson and John C. Gallant.

Filtering based on aspect

Let's satisfy the first customer by showing the houses with correct aspect values. First of all, we need the **SRTM DEM** we downloaded and clipped to our study area, then transformed to our projection. If we load it, we can use GDAL's raster processing algorithms to do some terrain analysis.

As QGIS does not have many tools dedicated to raster analysis, it uses external modules. This implies one very important specificity of raster analysis in QGIS--we cannot use memory layers. We have to save every intermediate result to disk. This is the case in calculating the aspect of our DEM, which you can do as follows:

1. Open **Raster | Terrain Analysis | Aspect** from the menu bar.
2. Select the clipped SRTM DEM as **Elevation layer**.
3. Specify an output, preferably in your working folder where you saved the clipped SRTM layer.
4. Click on **OK**:

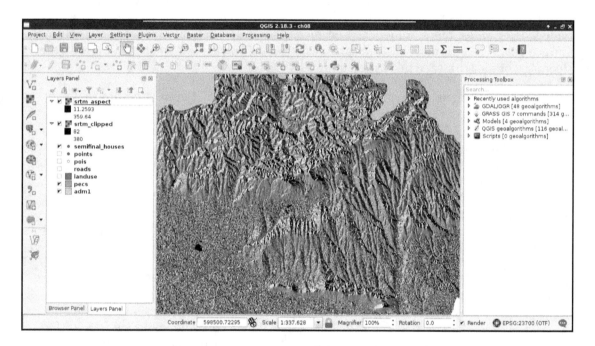

 If you do not have **Terrain Analysis** in the **Raster** menu, you might have to enable the containing plugin first. You can do this from the **Plugins | Manage and Install Plugins** dialog by enabling the **Raster Terrain Analysis plugin**.

Now we have an aspect layer, which resembles the hillshading we used in a previous chapter. The only, and very significant difference is, that we did not calculate shadows cast to the surface by a light source from a specific direction, but the surface's absolute exposure. Before going on with the analysis, let's interpret the result. Values of the aspect map range from near 0 to near 360. This corresponds to exposure values expressed in degrees.

More importantly, we must be able to map aspect values to directions. This sounds trivial; however, start directions can change from GIS to GIS. In QGIS, 0° and 360° correspond to North. Therefore, as we go clockwise (imagine a compass), 90° is East, 180° is South, and 270° is West. If we zoom in, and toggle the visibility of the underlying DEM, we can also see some areas without any values. There is a special value in aspect maps--flat. Flat surfaces can be denoted with a special value, such as -1, or like in our case, can be defined as NULL. We have to consider these flat values when we filter our points:

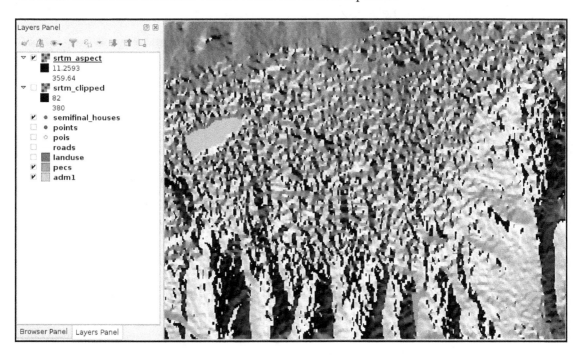

There is only one thing left--we have to sample the aspect map in the locations of our houses. Unfortunately, QGIS does not have a tool to achieve this task, but we can use GRASS GIS, which is really strong in raster analysis. Using GRASS has only one inconvenience--it won't create a column for the aspect values automatically:

1. Use **Field Calculator** on the filtered house layer, and add an empty field named `aspect` with a **Decimal number (real)** type. The precision value should be 3, and the expression can be a simple 0. Save the edits.
2. Open the **GRASS GIS 7 commands | Vector | v.what.rast.points** tool.
3. Select the houses as the vector points map.
4. Select the aspect layer as the raster map, which should be sampled.
5. Select the `aspect` column as the column to be updated.

6. Supply an expression which selects every feature, such as `id > 0`. Note that in GRASS, we do not enclose column names in quotation marks.

7. Specify an output in the **Sampled** field.

If GRASS complains about a projection mismatch, it means that you used the SRTM which was clipped, but not transformed. You can either transform the aspect map with **Save As**, or open the transformed SRTM, and calculate the aspect again.

Now we have the houses with the sampled aspect values. Therefore, we can select the final set of houses with an expression, such as the following:

```
"aspect" < 270 AND "aspect" > 90
```

Or, if we would like to be more restrictive, we can use this:

```
"aspect" < 225 AND "aspect" > 135
```

You might have a bad character encoding in GRASS GIS's output file. If this is the case, you can join the `aspect` column to the original filtered houses layer based on their `id` columns. Considering this case, it is recommended to use vector layers only with unique numeric columns with GRASS GIS. Furthermore, it is also recommended not to use any non-ASCII characters or whitespaces in path names when using GRASS.

Calculating walking times

The final preference was that the house should be within a 15 minutes walking distance to the train and bus stations. We could calculate the time taken to reach a point from another one on a road network using network analysis. However, that method would not respect an important factor in walking--elevation. This is a kind of nontrivial analysis done on a DEM. Although neither QGIS, nor PostGIS offer a solution for this type of analysis, GRASS GIS has just the tool for us. This tool creates a cost surface, which represents walking time in seconds from one or more input points. For this, GRASS needs a DEM and an additional friction map. With the friction map, we can fine-tune the analysis, giving weights to some of the areas. For example, we can give a very high value to buildings, making GRASS think that it's almost impossible to walk through them. The friction map must be in the raster format, therefore, our first task is to create it from our vectors.

For the sake of simplicity, let's only consider buildings, forests, and parks.

1. Apply a filter on the `landuse` layer to only show forests and parks. The correct expression is `"fclass" = 'park' OR "fclass" = 'forest'`.

2. Clip the `landuse` layer to the town's boundary. You can use memory layers if not stated otherwise.

3. Load the `buildings` layer, and clip it to the town's boundary. Save the clipped layer as a temporal layer, or to the disk.

4. Calculate the difference of the `landuse` layer and the `buildings` layer (in this order). It is convenient to not have overlapping or duplicated features in the final result. By calculating the difference of the two layers, we basically erase the `buildings` from the `landuse` layer. Save this layer as a temporal layer, or to the disk.

5. Merge the buildings and the difference layer with **QGIS geoalgorithms** | **Vector general tools** | **Merge vector layers**. You might want to build spatial indices on the input layers before (**Properties** | **General** | **Create spatial index**). Save the merged layer to the disk.

Why did we use **Merge vector layers** instead of **Union**? Try calculating the union of the two input layers. How many features do the input layers have? How many features does the union layer have? It should have as many features as the input layers. This is one of the many pitfalls of floating point arithmetic. The default behavior of the **Union** tool is cutting the overlapping parts of the inputs, and creating new features for them. In floating point arithmetic there is usually some rounding involved, and as a result, it is very hard to distinguish between overlapping and adjacent segments.

6. Give friction costs to the features using the **Field Calculator**. The field name should be `cost`, while the field type should be **Whole number (integer)**. Friction costs represent the penalty in seconds for crossing a single meter on the surface. For `buildings` layer, this penalty should be an arbitrary high number, such as 9000. For parks, the penalty can be 1, while for forests, the penalty can be 2. We can assign these costs conditionally using the following expression:

```
CASE
WHEN "fclass" = 'building' THEN 9000
WHEN "fclass" = 'forest' THEN 2
WHEN "fclass" = 'park' THEN 1
END
```

7. Save the edits and clean up (remove intermediary data):

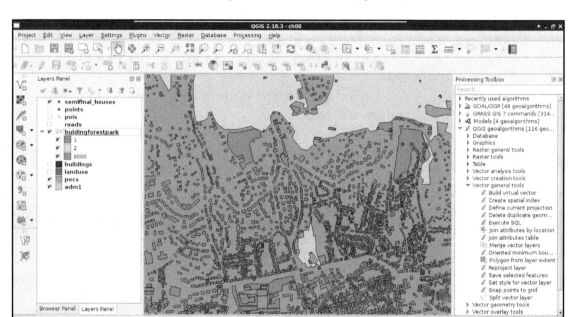

Now we have a vector friction layer, although GRASS needs a raster layer containing penalties. To transform our vector to raster, we can use GDAL's **Rasterize** module from **Raster | Conversion**. This module can create rasters from input vectors, although we should keep the differences of the two data models in mind:

- Rasters usually hold a single attribute, while vectors hold many. We have to choose a single attribute to work with.
- Rasters have a fixed resolution, while the concept of resolution does not apply to vector data, at least in this literal sense. We will lose data, and we have to choose how much we would like to lose. By choosing a very low spatial resolution, we can make our calculations very slow, while, if we choose too high value, we can get a faulty result (*Appendix 1.10*).

- As rasters consist of coincident cells, there is no guarantee the vector layer's bounds will match the resulting raster layer's bounds. Furthermore, if we use raster (matrix) algebra, there is no guarantee that the rasterized layer will fully overlap with other inputs. GRASS and GDAL handle these cases really well, although this is not universally true for every GIS.

Now that we have a concept about vector to raster transformation, let's create our friction map as follows:

1. Open the **Raster** | **Conversion** | **Rasterize** tool from the menu bar.
2. Choose the friction vector layer as an input, and the `cost` column as **Attribute field**.
3. Browse an output. Ignore the warning about creating a new raster. GDAL will handle that just right.
4. Choose the **Raster resolution in map units per pixel** radio box.
5. Provide 2 (or the equivalent if you use a unit other than meters) in both the **Horizontal** and **Vertical** fields.

Rasters produced by GDAL can be huge. In order to reduce the file size, you can click on the pencil icon before running the tool, and insert `-co "COMPRESS=DEFLATE" -ot Int16` right after the `gdal_rasterize` command.

6. The result is most likely a sole black raster. If you would like to see the different values, use a single band styling mode (**Properties** | **Style**), then specify the 0, 1, 2, and 9000 values as intervals:

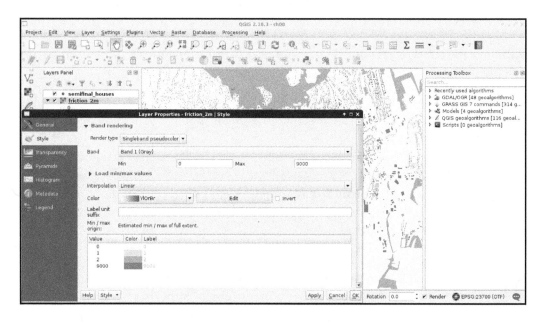

7. As the new raster map has most likely a custom projection with the same parameters as our local projection, assign the local projection to it instead. Otherwise, GRASS will complain about the projection mismatch. We can select our local projection in **Properties** | **General** | **Coordinate reference system**:

Besides the friction map, GRASS also needs an elevation map. However, we have some work to do on our SRTM DEM. First, as we have two raster layers with two resolutions, and it is no trivial matter for GRASS which one to choose, we have to resample our DEM to 2 meters. We should also clip the elevation raster to our town's boundary in order to reduce the required amount of calculations.

1. Open the SRTM DEM transformed to our local projection.
2. Select the **Raster** | **Extraction** | **Clipper** tool.
3. Select the DEM as the input layer, and browse an output layer.
4. Check the **No data value** box, thus, rasters outside of the town's boundary get NULL values.
5. Choose the **Mask layer** radio button.
6. Specify the town's boundary as a mask layer.

7. Check the **Crop the extent of the target dataset to the extent of the cutline** box, and run the tool.

8. Open the **Raster | Align Rasters** tool.

9. With the plus icon, add the clipped DEM to the raster list, and specify an output.

10. Check the **Cell Size** box, and provide the same cell size that the friction layer uses.

11. Run the tool.

The final parameter we should provide is a vector layer with a number of points representing starting points. We need two cost layers, one for the bus stations, and one for the railway stations. These data can be accessed from the `transport` OSM layer we inserted into PostGIS. Let's load that layer, and clip it to the town's boundary. We can store the result in a memory layer. Now we are only a few steps away from the cost surfaces:

1. Filter the clipped `transport` layer to only show railway stations first with the expression `"fclass" = 'railway_station'`.

2. Examine the railway stations. Are there any local stations which are irrelevant for our analysis? If there are, delete those points. Start an edit session with **Toggle Editing**, select the **Select Features by area or single click** tool, select an irrelevant station, then click on the **Delete Selected** button. After every local station is removed, save the edits and exit the edit session.

3. Are there any stations inside buildings? If there are, move them out, as they will produce incorrect results. Start an edit session, select the **Move feature(s)** tool, and move the problematic points outside of the buildings. Finally, save the edits, and exit the edit session.

4. Create the cost surface with the **GRASS GIS 7 commands | Raster | r.walk.points** tool. The input elevation map should be the SRTM layer clipped to the town's boundary, the input raster layer containing friction costs is our rasterized friction layer, and the start points is the filtered, edited transport layer. All other parameters can be left with their default values.

The algorithm produces two maps, one for the costs and one for directions. The direction map is irrelevant for us; we don't have to load it into QGIS after the algorithm finishes. Furthermore, if you expand the **Advanced parameters** menu, you can check the **Use the 'Knight's move'** box for more accurate results. With this option, the algorithm also considers cells reachable by a knight's move (like on a chess table) from each cell besides the direct neighbors. For quicker results, also increase the maximum memory usable by the algorithm.

If we style the result, we can see how much time it takes to reach our houses from the railway stations. Now we have to repeat the previous steps with bus stations. To filter bus stations, we can use the expression `"fclass" = 'bus_station'`:

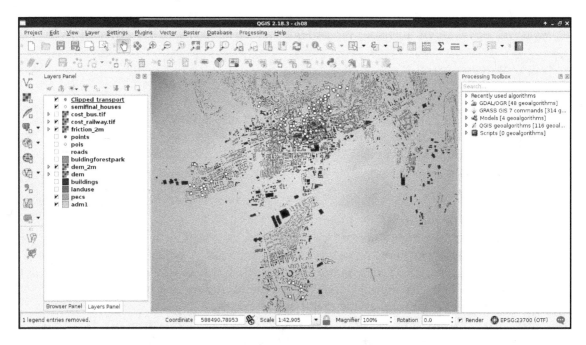

Almost there. We are only one step away from the final result--we need to sample both the cost maps at the houses' locations. Let's do this by following the steps we took earlier in this chapter:

1. Create two new fields for the houses layer with the names `railway` and `bus` using the **Field Calculator**. The precision and the type do not really matter in this case, as the costs are in seconds. They must have a numeric type though. The expression can be a single 0. Don't forget to save the edits, and exit the edit session once you have finished.

2. Use the **GRASS GIS 7 commands** | **Vector** | **v.what.rast.points** tool for sampling the first raster layer. Use a general expression, like `id > 0`, and save the result as a temporary layer.

3. Set the CRS of the output layer to the currently used projection in **Properties** | **General** | **Coordinate reference system**. GRASS applied its own definition of the same projection to the output, and will complain if it thinks that the projections of the input layers do not match.

4. Use the same tool again to sample the second raster. The input layer this time must be the output of the previous step. Save this result to the working folder.

5. Select preferred locations from the output with an expression. As the values are in seconds, we can build the expression `"railway" / 60 <= 15 AND "bus" / 60 <= 15` to select the relevant houses. If there are no matches, try to use a logical `OR` between the two queries:

Summary

Congratulations! You just carried out your first vector analysis. It was a great adventure, wasn't it? In this chapter, you learned how to utilize vector layers to derive some valuable results from them. We learned about some of the more popular and useful vector analysis tools, and ventured into the more advanced realm of GIS. We also learned how you can make your workflow more flexible by building models from your steps.

In the next chapter, we will talk about spatial analysis in PostGIS. We will see how we can build queries executing spatial operations on our tables, creating the same results as in this chapter by nesting queries in QGIS's database manager.

9
Spatial Analysis on Steroids - Using PostGIS

In the last chapter, we did some vector analysis to get meaningful results from our data. We went through the steps of different types of vector analysis technique, and also built models to make our work more convenient. However, the execution of the algorithms still took some time, and you might wonder if there is a more effective way for achieving the same results.

In this chapter, we will explore the PostGIS way of vector analysis, which is faster and more flexible, as spatial operations executed with SQL queries can be saved and reproduced easily. With this approach, we can additionally stay in the realm of our spatial database, which means we do not have to worry about memory layers, temporary data, or saved layers scattered through our working folder. To make this approach comparable to the previous chapter, we will go through the same task, with the twist of doing everything in PostGIS, and using QGIS only as a thin client that is, using QGIS only for visualizing data.

In this chapter, we will cover the following topics:

- Vector analysis in PostGIS
- Raster queries in PostGIS
- PostGIS-specific techniques

Delimiting quiet houses

First of all, we should upload our random houses in our spatial database. In order to do this, let's open QGIS's database manager through **Database | DB Manager | DB Manager**. Connect to the PostGIS database with the role we have--write privileges--and import the houses layer with **Import layer/file**. Of course, we have to open our raw houses layer, as it is not opened in QGIS. The options should be the same as before, which are as follows:

- Use the spatial schema.
- The table name should be `houses`.
- Check the **Create spatial index** box at least, but checking the **Create single-part geometries instead of multi-part** box as well will do no harm either.
- We do not have to define SRID information, as the data is already present in our local projection.

When we are done, let's visualize the uploaded `houses` layer in QGIS, and remove the one saved to the disk. If everything is fine, we should keep the DB manager window open, as we will execute our SQL queries there:

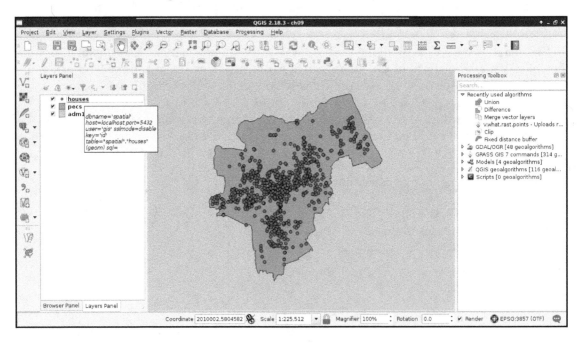

Proximity analysis in PostGIS

The first task in Chapter 8, *Spatial Analysis in QGIS* was to select every house which is at least 200 meters away from busy roads and 500 meters away from industrial areas. To complete this task, we created buffer zones around the problematic features and areas, calculated their union, and selected every disjoint house. Following the same approach, we can use two functions of PostGIS--ST_Intersects and ST_Buffer. We do not have to discuss ST_Intersects again, as we have already used it several times. ST_Buffer, on the other hand, is a different function. It does not act as a filtering function, but it creates and returns new geometries based on the input geometries and buffer distances. To buffer the busy roads of our roads layer, we can create the following expression:

```
SELECT ST_Buffer(geom, 200) AS geom
  FROM spatial.roads r
  WHERE r.fclass LIKE 'motorway%' OR r.fclass LIKE 'primary%';
```

If we execute the expression with the **Execute** button, we can see the resulting geometries, while, if we load the result as a layer by checking the **Load as a new layer** box, filling out the required fields, and selecting **Load now!**, we can see the buffered roads in QGIS:

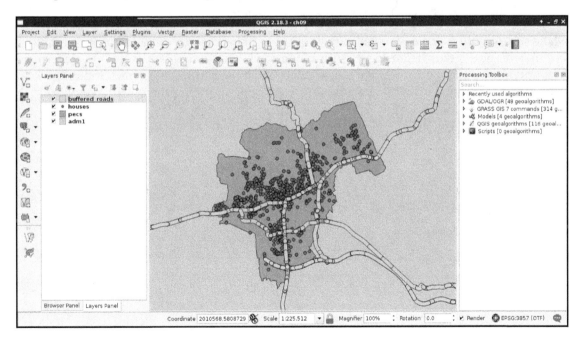

It is as easy as that. PostGIS iterates through the filtered roads, and buffers them on a row-by-row basis. It does not dissolve the results, or do anything else, but just returns the raw buffered geometries, and additionally, the attributes if we ask for them. As we will only use the geometries of the roads, we do not need any other attributes.

As PostGIS executes spatial functions on a row-by-row basis, we can create variable-sized buffer zones by specifying a numeric attribute field instead of the constant number as the second argument of ST_Buffer.

The only thing left to do is to check if our houses intersect with the buffer zones. As we only need a subset of the roads, which is effectively a subquery, we should precalculate the buffer zones in a **CTE (Common Table Expression)** table by using the WITH keyword as follows:

```
WITH main_roads AS (
  SELECT ST_Buffer(geom, 200) AS geom
  FROM spatial.roads r
  WHERE r.fclass LIKE 'primary%' OR r.fclass LIKE 'motorway%')
SELECT h.* FROM spatial.houses h, main_roads mr
  WHERE NOT ST_Intersects(h.geom, mr.geom);
```

Make sure to only execute the query; do not load as a layer. What did you get as a result? For me, this query returned more than 1 million rows. That's definitely a cross join between the individual buffer zones and the houses. Let's see if the results are correct, though. We can select only unique rows by using the DISTINCT operator after the SELECT keyword:

```
WITH main_roads AS (
  SELECT ST_Buffer(geom, 200) AS geom FROM spatial.roads r
  WHERE r.fclass LIKE 'primary%' OR r.fclass LIKE 'motorway%')
SELECT DISTINCT h.* FROM spatial.houses h, main_roads mr
  WHERE NOT ST_Intersects(h.geom, mr.geom);
```

The preceding query returns 1000 rows. That's the number of features we have in our houses layer.

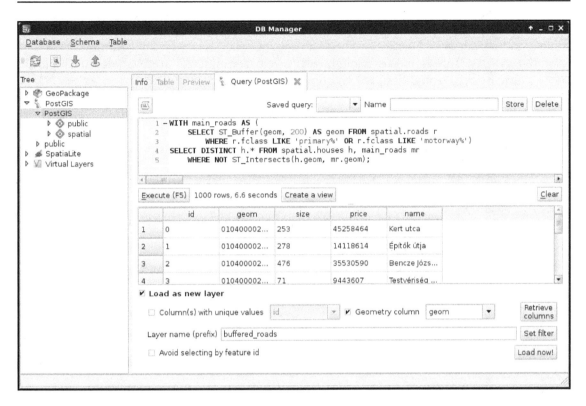

By creating a cross join, PostgreSQL returned every occurrence, where a house is disjoint with a buffer zone. That is, it returned every house multiple times. The magic we can use to solve this problem is to calculate the union of the buffer zones with ST_Union:

```
WITH main_roads AS (
  SELECT ST_Union(ST_Buffer(geom, 200)) AS geom
  FROM spatial.roads r
  WHERE r.fclass LIKE 'primary%' OR r.fclass LIKE 'motorway%')
SELECT h.* FROM spatial.houses h, main_roads mr
  WHERE NOT ST_Intersects(h.geom, mr.geom);
```

You must be wondering if this even makes any sense. If PostGIS goes row by row, why would it matter if we use union on every individual buffered geometry? The answer is that ST_Union is an aggregate function. Aggregate functions in PostGIS behave differently when they are called with one argument in a SELECT clause. They act on every returned row, and return a single geometry if there are no additional groupings defined. Now that the main_roads table has a single row, PostgreSQL behaves nicely, and returns only the disjoint features in less than a second.

The only thing left to do is to query houses which do not reside in the 500 meters buffer zone of industrial areas. For this, we need an additional CTE table with the union of the buffered industrial areas. In PostgreSQL, we can create multiple virtual tables in a single WITH clause by separating them with commas as follows:

```sql
WITH main_roads AS (
  SELECT ST_Union(ST_Buffer(geom, 200)) AS geom
  FROM spatial.roads r
  WHERE r.fclass LIKE 'primary%' OR r.fclass LIKE 'motorway%'),
  industrial_areas AS (
   SELECT ST_Union(ST_Buffer(geom, 500)) AS geom
   FROM spatial.landuse l
   WHERE l.fclass = 'industrial' OR l.fclass = 'quarry')
SELECT h.* FROM spatial.houses h, main_roads mr,
  industrial_areas ia
  WHERE NOT ST_Intersects(h.geom, mr.geom) AND NOT
  ST_Intersects(h.geom, ia.geom);
```

Quite a complex query, isn't it? On the other hand, it still returns the correct houses in less than a second, which is a serious performance boost. Let's load the results as a layer in QGIS, named something like quiet_houses_buffer:

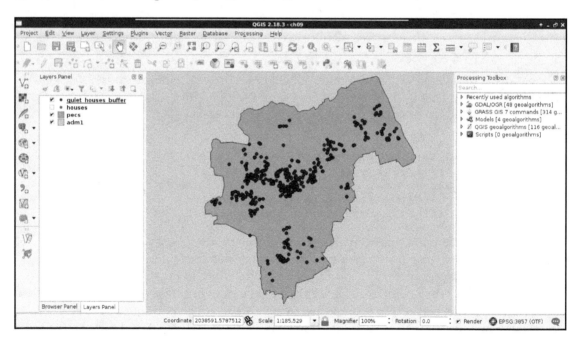

What is really great about PostGIS is that we do not have to rely on traditional ways of analysis. We can cook up new methods if they can be done on a row-by-row basis, and return correct results. For example, we do not have to buffer the constraint features, just measure their distances to the houses with ST_Distance. This way, we can use a more lightweight aggregate function--ST_Collect. It does not calculate the union of the input geometries, but just merges them into a single one. If we supply polygons, the resulting geometry will be a multipart polygon containing the input polygons as members.

> Using ST_Collect with ST_Intersects does not work out well every time. It is more reliable to use ST_Union with ST_Intersects.

Let's create a query using ST_Distance and ST_Collect:

```
WITH main_roads AS (
  SELECT ST_Collect(geom) AS geom
  FROM spatial.roads r
  WHERE r.fclass LIKE 'primary%' OR r.fclass LIKE 'motorway%'),
industrial_areas AS (
  SELECT ST_Collect(geom) AS geom FROM spatial.landuse l
  WHERE l.fclass = 'industrial' OR l.fclass = 'quarry')
SELECT h.* FROM spatial.houses h, main_roads mr,
  industrial_areas ia
  WHERE ST_Distance(h.geom, mr.geom) > 200 AND ST_Distance(
  h.geom, ia.geom) > 500;
```

This method has a similar performance to the last one; however, it returned fewer features for me. If you experienced the same, load the results of this query in QGIS, and examine the difference.

Precision problems of buffering

If you also have some differences between the buffering and the distance measurement methods, follow me during the investigation. If not, just read along. First, let's see the problematic geometries in context. I only have one; therefore, it is more trivial to find out the source of the problem:

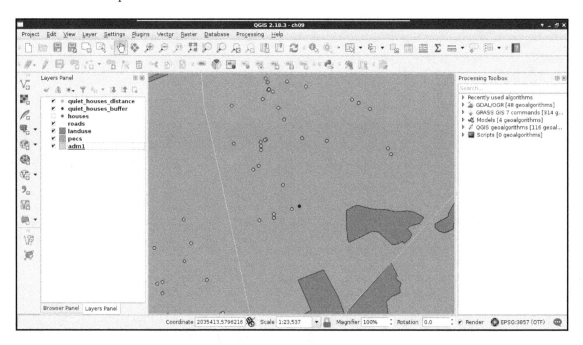

The main roads are far enough from the feature; therefore, the nearest industrial area should be the cause. If we create a 500-meters buffer zone around the layer, we should see that the feature lies outside of the buffer zone. However, `ST_Distance` thought otherwise. What we can also see is that the buffer zones are not circular; they have sides approximating the real buffer zones. The problem is that buffers are regular polygons, therefore, they consist of vertices connected by straight segments. That is, we cannot create true circular buffers. There is no problem when we buffer straight lines, the result becomes exact. The precision problem only arises when we buffer corners, and the buffer zones should be circular. We can, however, create better approximations by using more vertices for creating these circular arcs. In QGIS, we can use the **Segments** option in the buffer tools to specify the level of detail, and increase the precision of the result (*Appendix 1.11*).

PostGIS also allows us to specify the number of segments used for approximating a quarter circle. We can provide a third parameter to the ST_Buffer function, which represents the number of segments. Let's modify the query to use 25 segments for the approximation as follows:

```
WITH main_roads AS (
  SELECT ST_Union(ST_Buffer(geom, 200, 25)) AS geom
  FROM spatial.roads r
  WHERE r.fclass LIKE 'primary%' OR r.fclass LIKE 'motorway%'),
industrial_areas AS (
  SELECT ST_Union(ST_Buffer(geom, 500, 25)) AS geom
  FROM spatial.landuse l
  WHERE l.fclass = 'industrial' OR l.fclass = 'quarry')
SELECT h.* FROM spatial.houses h, main_roads mr,
  industrial_areas ia
  WHERE NOT ST_Intersects(h.geom, mr.geom) AND NOT
  ST_Intersects(h.geom, ia.geom);
```

Now the results match the ones from using exact minimum distance measurements, although the query slowed down a bit, making the other query the obvious choice.

Querying distances effectively

While using intersection checks with buffer zones only yields Boolean results (inside or outside), calculating the minimum distance from the reference geometry holds other advantages. For example, our customer can ask which of the houses are the farthest from those noisy areas. By querying the distances, we can easily answer that question. We can even order our results by the combined distances using ORDER BY at the end of our query. Let's remove the buffer-based query's layer, and modify the distance-based one's expression by right-clicking on it, and selecting **Update Sql Layer**:

```
WITH main_roads AS (
  SELECT ST_Collect(geom) AS geom
  FROM spatial.roads r
  WHERE r.fclass LIKE 'primary%' OR r.fclass LIKE 'motorway%'),
industrial_areas AS (
  SELECT ST_Collect(geom) AS geom
  FROM spatial.landuse l
  WHERE l.fclass = 'industrial' OR l.fclass = 'quarry')
SELECT h.*, ST_Distance(h.geom, mr.geom) AS dist_road,
  ST_Distance(h.geom, ia.geom) AS dist_ind
  FROM spatial.houses h, main_roads mr, industrial_areas ia
  WHERE dist_road > 200 AND dist_ind > 500
ORDER BY dist_road + dist_ind DESC;
```

In theory, we are now asking PostgreSQL to evaluate the two distance checks, and return them to us in the `dist_road` and `dist_ind` columns. Then we simply use those columns to select only the correct houses, and order the results in descending order.

> If you would like to order the results in ascending order, you do not have to specify anything after the ORDER BY statement, as that is the default behavior in PostgreSQL. Ordering based on the sum of the two distance columns in ascending order would look like ORDER BY dist_road + dist_ind.

What happens if we run this query? It returns an error with the message `"dist_road"` `does not exists`. We tried to use a field which had not yet been calculated when PostgreSQL tried to call it. We simply cannot use computed fields in other queries on the same level. Let's modify our query a little bit, as follows:

```
WITH main_roads AS (
 SELECT ST_Collect(geom) AS geom
 FROM spatial.roads r
 WHERE r.fclass LIKE 'primary%' OR r.fclass LIKE 'motorway%'),
industrial_areas AS (
 SELECT ST_Collect(geom) AS geom
 FROM spatial.landuse l
 WHERE l.fclass = 'industrial' OR l.fclass = 'quarry')
SELECT h.*, ST_Distance(h.geom, mr.geom) AS dist_road,
 ST_Distance(h.geom, ia.geom) AS dist_ind
 FROM spatial.houses h, main_roads mr, industrial_areas ia
 WHERE ST_Distance(h.geom, mr.geom) > 200 AND ST_Distance(h.geom,
 ia.geom) > 500
ORDER BY ST_Distance(h.geom, mr.geom) + ST_Distance(h.geom,
 ia.geom) DESC;
```

The new query works, although the distances are calculated three times, slowing down the execution, which is bad. On the other hand, we have the distance values, and our features are ordered based on their combined distances from noisy areas:

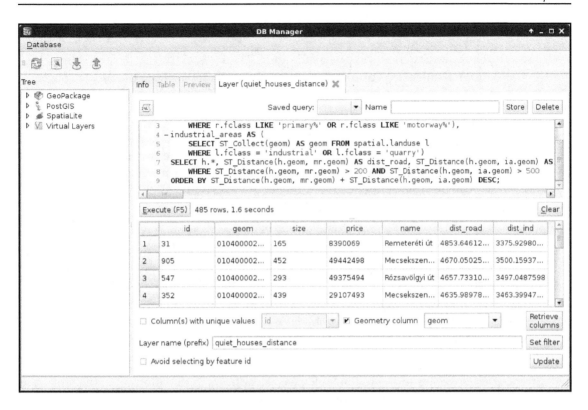

We should still do something about this query, as it is quite inconvenient to declare the same calculations multiple times. If we wish to change some of the distance columns, we would have to change them in three places. This is the case when a subquery proves useful. We can calculate the `houses` table with the distance columns as a new table called `hwd` (houses with distances). Then we can simply refer to the columns of that table in the other two occurrences:

```
WITH main_roads AS (
 SELECT ST_Collect(geom) AS geom
 FROM spatial.roads r
 WHERE r.fclass LIKE 'primary%' OR r.fclass LIKE 'motorway%'),
industrial_areas AS (
 SELECT ST_Collect(geom) AS geom
 FROM spatial.landuse l
 WHERE l.fclass = 'industrial' OR l.fclass = 'quarry')
SELECT * FROM
 (SELECT h.*, ST_Distance(h.geom, mr.geom) AS dist_road,
 ST_Distance(h.geom, ia.geom) AS dist_ind
 FROM spatial.houses h, main_roads mr, industrial_areas ia) AS hwd
 WHERE hwd.dist_road > 200 AND hwd.dist_ind > 500
```

```
ORDER BY hwd.dist_road + hwd.dist_ind DESC;
```

Using subqueries and precalculating CTE tables in the WITH clause are interchangeable. You can also include the hwd table along the main_roads and industrial_areas tables for better readability.

The calculation time is the same, as PostgreSQL has to query the distances for every house before it can query the relevant rows. Additionally, using subqueries makes the code harder to read and interpret. Luckily, PostgreSQL offers us an even better method for creating columns, which can be cross-referenced--lateral subqueries. By using the keyword LATERAL before a subquery, we only need to include the formulae for the dynamically calculated columns, and we can use them outside of the subquery.

The LATERAL keyword can be used before subqueries and joins in PostgreSQL. It is basically a special cross join with computed columns. The magic part is that this method leaves plenty of space for PostgreSQL to optimize our queries. You can read more about lateral joins at https ://www.periscopedata.com/blog/reuse-calculations-in-the-same-q uery-with-lateral-joins.html, and by following the links provided there.

Lateral subqueries can be added to the tables we select from, and, therefore, in the FROM clause. We only need to provide the computed columns and their formulae in the subquery as follows:

```
WITH main_roads AS (
  SELECT ST_Collect(geom) AS geom FROM spatial.roads r
  WHERE r.fclass LIKE 'primary%' OR r.fclass LIKE 'motorway%'),
industrial_areas AS (
  SELECT ST_Collect(geom) AS geom FROM spatial.landuse l
  WHERE l.fclass = 'industrial' OR l.fclass = 'quarry')
SELECT h.*, hwd.dist_road, hwd.dist_ind
  FROM spatial.houses h, main_roads mr, industrial_areas ia,
  LATERAL (SELECT ST_Distance(h.geom, mr.geom) AS dist_road,
  ST_Distance(h.geom, ia.geom) AS dist_ind) AS hwd
  WHERE hwd.dist_road > 200 AND hwd.dist_ind > 500
ORDER BY hwd.dist_road + hwd.dist_ind DESC;
```

With this query, we not only made the code more readable, but also increased its performance by almost 30%. As PostgreSQL could optimize our new lateral query better, we get similar performance as with the precise buffering technique, acquired the distances from the noisy areas, and ordered our features according to them. With this final optimization, there is no more doubt that distance calculations should be preferred over buffering in PostGIS for proximity analysis.

Saving the results

Finally, let's save the last query as a view without ordering the results. This way, we can ease our future work by having shorter, and thus, more manageable chunks of code. We can easily create a view from QGIS's **DB Manager** by prefixing the query with `CREATE VIEW AS` and the view's name. However, before creating a view, let's talk about naming conventions. If we work with a PostGIS database extensively, we will definitely end up with different kinds of tables like the following:

- Original tables holding raw spatial data for our analyses, which shouldn't be modified or dropped accidentally
- Tables holding final or partial results of our analyses
- Views and materialized views holding queries, and speeding up our work

To distinguish between the different types of data, we should apply a naming convention. A good example would be prefixing different kinds of data with some abbreviations. This practice can help pgAdmin or other graphical interfaces to visually group and organize different kinds of data. We can name our views as `vw_viewname` or `vwViewName`, our temporal or final results as `res_tableName` or `resTableName`, and so on. It shouldn't really matter (although PostgreSQL will drop camel cases by default) as long as we keep to our rules, and name our tables consistently. Now as I'm naming my view `vw_quiethouses`, the query saving the view looks like the following:

```
CREATE VIEW spatial.vw_quiethouses AS WITH main_roads AS (
  SELECT ST_Collect(geom) AS geom FROM spatial.roads r
  WHERE r.fclass LIKE 'primary%' OR r.fclass LIKE 'motorway%'),
industrial_areas AS (
  SELECT ST_Collect(geom) AS geom FROM spatial.landuse l
  WHERE l.fclass = 'industrial' OR l.fclass = 'quarry')
SELECT h.*, hwd.dist_road, hwd.dist_ind
  FROM spatial.houses h, main_roads mr, industrial_areas ia,
  LATERAL (SELECT ST_Distance(h.geom, mr.geom) AS dist_road,
  ST_Distance(h.geom, ia.geom) AS dist_ind) AS hwd
WHERE hwd.dist_road > 200 AND hwd.dist_ind > 500;
```

You can also create a view from QGIS by opening a new **SQL window**, typing or copying the query, and clicking on the **Create a view** button. However, views created this way are automatically put into the **public** schema (or the default one), no matter whether you use a qualified name (that is, include the name of the destination schema).

Matching the rest of the criteria

We have two additional tasks to complete in order to match the common preferences of our customers. If you do not remember the exact preferences, here they are again:

- They should be less than 500 meters away from a park with a playground
- They should be less than 500 meters away from a restaurant
- They should be less than 500 meters away from a bar or a pub
- There should be at least two markets within their 500 meters vicinity

The first three criteria can be easily matched building on the queries of the previous section. We only have to create three CTE tables or subqueries; one for the parks with playgrounds in their 200 meters vicinity, one for the restaurants, and one for the bars and pubs. After that, we only have to match our houses by using distance checks. Such a query can be formulated as follows:

```
WITH parks_with_playgrounds AS (
  SELECT ST_Collect(l.geom) AS geom FROM spatial.landuse l,
  spatial.pois p
  WHERE l.fclass = 'park' AND p.fclass = 'playground' AND
ST_DWithin(l.geom, p.geom, 200)),
restaurants AS (
  SELECT ST_Collect(geom) AS geom FROM spatial.pois p
  WHERE p.fclass = 'restaurant'),
bars AS (
  SELECT ST_Collect(geom) AS geom FROM spatial.pois p
  WHERE p.fclass = 'bar' OR p.fclass = 'pub')
SELECT h.* FROM spatial.vw_quiethouses h,
  parks_with_playgrounds pwp, restaurants r, bars b
  WHERE ST_DWithin(h.geom, pwp.geom, 500) AND ST_DWithin(h.geom,
    r.geom, 500) AND ST_DWithin(h.geom, b.geom, 500);
```

The final expression works as the ones formulated before. We filter the raw tables by attributes and distances, and use some distance checks in the final query. The only difference is that as we do not need the exact distances, we used ST_DWithin instead of ST_Distance. ST_DWithin is a simple filtering function for selecting features based on a proximity. It has almost no performance penalty over ST_Distance, but offers a more concise way to formulate our query. It needs three arguments: two geometries, and a distance in the units of the input layers' SRID.

 As subqueries are evaluated independently, you can use the same aliases in different subqueries without confusing PostgreSQL. You just have to make sure that you do not confuse yourself.

If we load the result as a layer, we can see our constrained quiet houses, which match the results created in QGIS at a first glance:

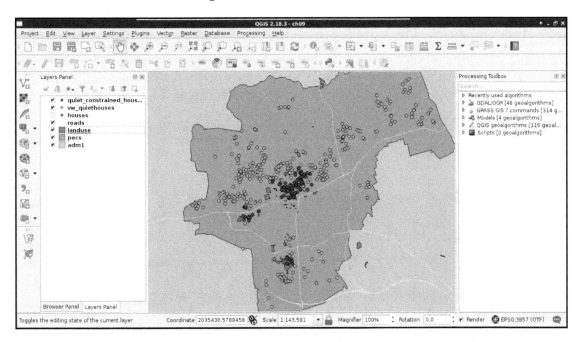

For the sake of simplicity and clarity, let's create another view from the result of this query. We just have to prefix our query with CREATE VIEW AS and the view's name:

```
CREATE VIEW spatial.vw_quietconstrainedhouses AS WITH
parks_with_playgrounds AS (
SELECT ST_Collect(l.geom) AS geom FROM spatial.landuse l,
spatial.pois p
WHERE l.fclass = 'park' AND p.fclass = 'playground' AND
ST_DWithin(l.geom, p.geom, 200)),
restaurants AS (
SELECT ST_Collect(geom) AS geom FROM spatial.pois p
WHERE p.fclass = 'restaurant'),
bars AS (
SELECT ST_Collect(geom) AS geom FROM spatial.pois p
WHERE p.fclass = 'bar' OR p.fclass = 'pub')
SELECT h.* FROM spatial.vw_quiethouses h,
parks_with_playgrounds pwp, restaurants r, bars b
WHERE ST_DWithin(h.geom, pwp.geom, 500) AND ST_DWithin(h.geom,
r.geom, 500) AND ST_DWithin(h.geom, b.geom, 500);
```

Counting nearby points

Although PostGIS is one of the state-of-the-art GIS softwares for effective spatial queries, aggregating effectively can sometimes be tricky. For this reason, we will go step by step through appealing to the final criterion. First of all, we need to select the markets from our POI table. This should be easy, as we just have to chain some queries on a single column together with the logical OR operator. Or we can use a more convenient operator created for similar tasks called IN. By using IN, we can supply a collection of values to check a single column against:

```
SELECT geom FROM spatial.pois p
  WHERE p.fclass IN ('supermarket', 'convenience',
  'mall', 'general');
```

Let's put this table in a WITH clause, and go on with counting points.

```
WITH markets AS (
  SELECT geom FROM spatial.pois p
  WHERE p.fclass IN ('supermarket', 'convenience',
  'mall', 'general'));
```

We used some aggregating functions before, but only for the sole purpose of returning a single aggregated value for an entire table. Aggregating a single column is trivial enough for PostgreSQL not to ask for any other parameters. However, if we would like to select multiple columns while aggregating, we have to specify our intention of creating groups explicitly by using a GROUP BY expression with a selected column name. We can create a simple grouping by querying the IDs of our houses layer along with the number of geometries in our markets layer. Counting can be done by utilizing count, one of the most basic aggregating function in PostgreSQL, as follows:

```
WITH markets AS (
  SELECT geom FROM spatial.pois p WHERE p.fclass IN ('supermarket',
  'convenience', 'mall', 'general'))
SELECT h.id, count(m.geom) AS count FROM spatial.houses h,
  markets m
  GROUP BY h.id;
```

Now we got 1,000 results, just as many houses we have. Every row has the total count of geometries in our markets layer, as we did not supply a condition for the selection. Basically, we got the cross join of the two tables, but grouped by the IDs of our houses table. As we aggregated the number of geometries in each group, and got every possible combination, we ended up with the same number of points for every group.

By separating multiple tables in the FROM clause with commas, technically, PostgreSQL applies an inner join. On the other hand, as an inner join is a subset of the cross join matching the join conditions, we ended up with the cross join by not specifying any conditions.

Let's supply a join condition to narrow down our results:

```
WITH markets AS (
 SELECT geom FROM spatial.pois p WHERE p.fclass IN ('supermarket',
 'convenience', 'mall', 'general'))
SELECT h.id, count(m.geom) AS count FROM spatial.houses h,
 markets m
WHERE ST_DWithin(h.geom, m.geom, 500)
GROUP BY h.id;
```

By providing the condition we should be able to see a subset of our data:

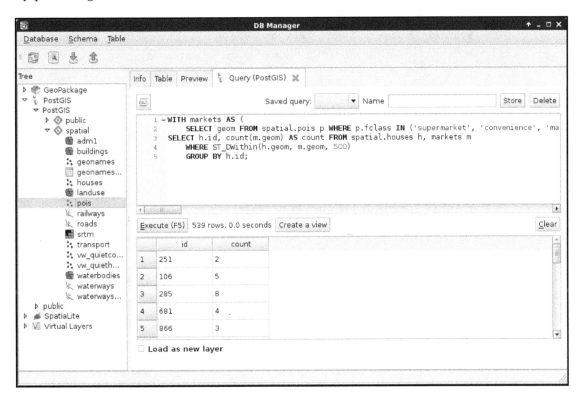

Now we get a real inner join; PostgreSQL only returned the rows matching the join conditions. That is, we got a table where every row with a geometry closer than 500 meters to a house got joined to it. From that table, PostgreSQL could easily aggregate the number of markets in the 500 meters vicinity of our houses. Note that we got back only a part of our tables, as features without markets (empty groups) got discarded. Let's take a note about that number, as we will need it later.

Before going further, let's rephrase this expression to include the INNER JOIN keywords. If we implicitly define a join, that is, separate layers in the FROM clause with commas, we can specify our join conditions in the WHERE clause. However, if we define a join explicitly, the conditions go in the ON clause:

```
WITH markets AS (
 SELECT geom FROM spatial.pois p
 WHERE p.fclass IN ('supermarket', 'convenience',
 'mall', 'general'))
SELECT h.id, count(m.geom) AS count
 FROM spatial.houses h INNER JOIN markets m
 ON ST_DWithin(h.geom, m.geom, 500)
 GROUP BY h.id;
```

Now we can easily change the join between our tables to an outer-left join in order to have the rest of the rows with 0 markets in their 500 meters radii:

```
WITH markets AS (
 SELECT geom FROM spatial.pois p
 WHERE p.fclass IN ('supermarket', 'convenience',
 'mall', 'general'))
SELECT h.id, count(m.geom) AS count
 FROM spatial.houses h LEFT JOIN markets m
 ON ST_DWithin(h.geom, m.geom, 500)
 GROUP BY h.id;
```

As the query's result shows, we have 1,000 rows, just like our houses table. Some of the rows have 0 values. However, are these results the same as the previous ones? We can do a quick validation by filtering out groups with 0 count values. If we get the same number of rows we noted previously, then we are probably on the right track.

To select from groups, we cannot use a WHERE clause; we have to use a special clause designed for filtering groups--HAVING:

```
WITH markets AS (
 SELECT geom FROM spatial.pois p
 WHERE p.fclass IN ('supermarket', 'convenience',
 'mall', 'general'))
SELECT h.id, count(m.geom) AS count
 FROM spatial.houses h LEFT JOIN markets m
 ON ST_DWithin(h.geom, m.geom, 500)
 GROUP BY h.id HAVING count(m.geom) > 0;
```

For me, PostgreSQL returned the same number of rows; however, using LEFT JOIN and filtering with a HAVING clause slowed down the query. Before creating a CTE table along with markets from the result, we should rewrite our count table's query to its previous, faster form:

```
WITH markets AS (
 SELECT geom FROM spatial.pois p
 WHERE p.fclass IN ('supermarket', 'convenience',
 'mall', 'general')),
 marketcount AS (SELECT h.id, count(m.geom) AS count
 FROM spatial.houses h, markets m
 WHERE ST_DWithin(h.geom, m.geom, 500)
 GROUP BY h.id);
```

Now the only thing left to do is to select the houses from our last view which have at least two markets in their vicinity:

```
WITH markets AS (
 SELECT geom FROM spatial.pois p
 WHERE p.fclass IN ('supermarket', 'convenience',
 'mall', 'general')),
 marketcount AS (SELECT h.id, count(m.geom) AS count
 FROM spatial.houses h, markets m
 WHERE ST_DWithin(h.geom, m.geom, 500)
 GROUP BY h.id)
SELECT h.* FROM spatial.vw_quietconstrainedhouses h,
 marketcount m
 WHERE h.id = m.id AND m.count >= 2;
```

By supplying the full query, we can see our semifinal results on our map:

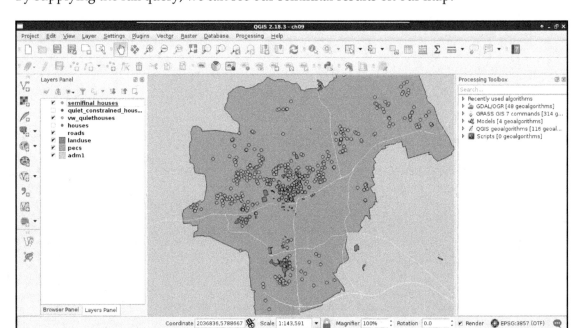

Look at that performance boost! For me, the whole analysis took about 1.3 seconds. On top of that, we can alter any parameter just by changing the view definitions. Additionally, we got the distances from the noisy places on which we can order our features. By ordering the result in a decreasing order, we can label our features according to that parameter, and show them to our customers on a map.

QGIS respects the order of the features coming from a PostGIS database by assigning a special _uid_ attribute column to them.

Finally, let's save our semifinal results as a third view:

```
CREATE VIEW spatial.vw_semifinalhouses AS WITH markets AS (
  SELECT geom FROM spatial.pois p
  WHERE p.fclass IN ('supermarket', 'convenience',
  'mall', 'general')),
  marketcount AS (SELECT h.id, count(m.geom) AS count
  FROM spatial.houses h, markets m
  WHERE ST_DWithin(h.geom, m.geom, 500)
  GROUP BY h.id)
  SELECT h.* FROM spatial.vw_quietconstrainedhouses h, marketcount m
```

```
WHERE h.id = m.id AND m.count >= 2;
```

Querying rasters

As PostGIS has limited raster capabilities compared to the sophisticated algorithms that GRASS GIS has, we have no way to calculate walking distances in our spatial database. However, in PostGIS, we can query raster tables and carry out basic terrain analysis, like calculating aspect. Querying raster layers with points is a surprisingly fast operation in PostGIS, as it can use the bounding boxes of raster tiles for geometry indexing, transform our points to pixel coordinates in the correct tile, and get the corresponding value from the stored binary raster by calculating an offset in bytes. We can use the ST_Value function to query raster data as follows:

```
SELECT h.*, ST_Value(r.rast, h.geom) AS elevation
  FROM spatial.vw_semifinalhouses h, spatial.srtm r
  WHERE ST_Intersects(r.rast, h.geom);
```

The only limitation of ST_Value is that it only accepts single-part points. Therefore, if we stored our houses as multipoint geometries, we need to extract the first geometry from them manually. If you got an error for the preceding query, that is a probable case. We can extract single-part geometries from a multipart geometry with the ST_GeometryN function, which needs a multipart geometry and a position as arguments. If we saved our houses table as multipoints, each geometry holds the single-part representation of our houses in its first position:

```
SELECT h.*, ST_Value(r.rast, ST_GeometryN(h.geom, 1)) AS elevation
  FROM spatial.vw_semifinalhouses h, spatial.srtm r
  WHERE ST_Intersects(r.rast, h.geom);
```

Although raster queries are fast in PostGIS, raster calculations are quite slow, as PostGIS has to execute the required operations on the requested tiles. There are a lot of possibilities from which we will use the ST_Aspect function to calculate the aspect in the locations of our houses. It is quite easy to add this function to our query, as it only needs a raster as an input. Furthermore, we should modify our query to only return houses with a southern aspect:

```
SELECT h.*, a.aspect FROM spatial.vw_semifinalhouses h,
  spatial.srtm r,
  LATERAL (SELECT ST_Value(ST_Aspect(r.rast), ST_GeometryN(h.geom,
  1)) AS aspect) AS a
  WHERE ST_Intersects(r.rast, h.geom) AND a.aspect < 225 AND
  a.aspect > 135;
```

You can find other raster-related functions in PostGIS's raster reference at `http://postgis.net/docs/RT_reference.html#Raster_Processing`.

Great work! We just fulfilled every criteria of one of our customers entirely in PostGIS. Although raster calculations are faster in QGIS and GRASS, and uploading rasters into PostGIS is cumbersome, it is worth considering uploading processed rasters to PostGIS for the convenience and performance of plain raster queries:

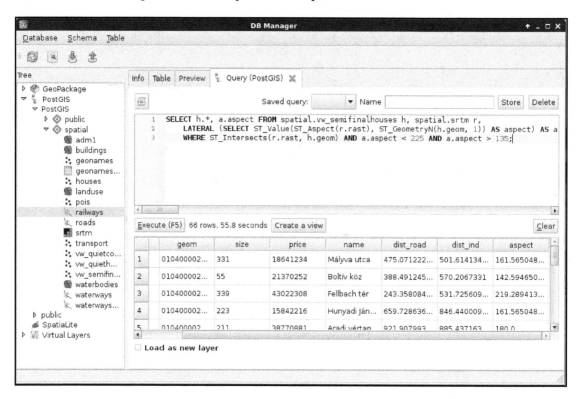

Summary

In this chapter, we learned how to improve the speed of our vector analysis by orders of magnitude. We simply used one of the state-of-the-art tools for quick vector analysis-- PostGIS. We also learned more about vector analysis, some of their pitfalls, and how to get more out of our spatial database. We carried out a spatial analysis, which would have been cumbersome in other desktop GIS software, to gain valuable extra information from our data. Of course, PostGIS and PostgreSQL have capabilities far beyond the scope of this chapter; therefore, if you are planning to work with spatial relational databases, it is definitely worth digging in deeper, and reading additional sources focused on PostGIS.

In the next chapter, we will focus on raster analysis, and learn about the most essential raster tools. We will create a decision problem where we have to choose the best site based on some criteria. Finally, we will not only solve that problem, but additionally, use some statistical methods to evaluate the result and create a clear basis for a well-founded decision.

10
A Typical GIS Problem

In the last chapter, we discussed vector analysis, and how we can perform it effectively. After explaining the basics in QGIS, we harnessed the power of PostGIS and carried out our analysis with unpaired speed. We also queried rasters, and executed a basic terrain analysis operation getting aspect values in the locations of our randomly generated houses as a result.

In this chapter, we will move on and discuss raster analysis in detail. We will learn how to use the most essential raster tools, and what kind of typical operations we can do with rasters. To spice up this chapter, first we create a scenario where we are decision makers. We search for the ideal site of our business, and we've already evaluated the criteria for the optimal site. The twist is that we are not looking for equally ideal sites, but searching for the best site for our purpose. Therefore, raw results showing possibilities are not enough in this case; we need an assessment evaluating those possibilities on which we can make our decision.

In this chapter, we will cover the following topics:

- Raster analysis
- Multi-criteria evaluation
- Fuzzy logic
- Basic statistics
- Creating an atlas

Outlining the problem

First of all, we need a scenario involving a problem for us to solve. In this chapter, we are decision makers looking for the best site for our business. We are supplying stores in multiple settlements scattered in our study area. More precisely, in every seat of the administrative division, we have stores to supply. We are looking for a site appropriate for holding our logistics center (that is, warehouse). As we will build the center, we do not need existing infrastructure on the site, although it should be economically feasible to build on it, and large enough to hold our 1 km^2 building with some loading area to load and unload supplies. For the sake of simplicity, the shape of the building is not important, we are flexible enough to conform to the chosen site. Last, but not least, we don't need a single site. We need a list of the most suitable sites from which we can choose the best one for our business. Summarizing and expanding the preferences, we can get a nice list of criteria as follows:

- The sites must be in our study area
- They should be as close as possible to every settlement we need to reach
- They should be as close to main roads as possible
- They should be empty, mostly flat, sites
- They should be large enough for the warehouse and the loading area

 An additional, very important factor for our analysis would be the type of the bedrock in the given site contributing to its stability. However, we neither have the required data for analyzing that feature, nor the scope for the theoretical background. For further reading, National Geographic's article at `http://www.nationalgeographic.org/encyclopedia/bedrock/` is a good starting point.

By translating these criteria to the language of our GIS model, we can create a more specific list, which is as follows:

- The validity extent of the analysis is our study area. We should only use data clipped to its bounds or clip the final result.
- The sites should be close to the mean point of the seats of administrative divisions in the study area. The closer, the better.
- They should be close to motorways and highways. Maximum 5 kilometers, but the closer, the better.
- They shouldn't overlap with forest areas, residential areas, industrial areas, and the like.

- The slope of the areas should be equal to or less than 10 degrees.
- The final areas should have at least an area of 1.5 km².

Additionally, to fully satisfy the preference of economical feasibility, we should add the following criterion:

- They shouldn't reside in the 200 meters vicinity of rivers and lakes, but the farther, the better.

 Why do we exclude areas in the close vicinity of rivers and lakes? To reduce the risk of damage caused by floods, of course. However, proper flood and floodplain analysis belongs to the domain of hydrology and hydrological modeling. Although QGIS and GDAL do not have tools for this discipline, you can take a look at GRASS's tools at `https://grasswik i.osgeo.org/wiki/Hydrological_Sciences`.

Raster analysis

Unlike our previous analysis, now that we do not have input points or areas to choose from, we have to delimit those areas based on different criteria. That alone raises the idea of using rasters. Additionally, this time we not only have Boolean criteria (inside or outside), but also have some continuous preferences (closer, or farther, the better). This factor calls for raster analysis. In raster analysis, we can consider almost the same classification as in vector analysis:

- **Overlay analysis**: Masking a raster layer with a binary mask layer. Where the binary mask layer has a zero value, we drop the value of the other raster layer, or set it to zero.
- **Proximity analysis**: Analyzing the distance between features or cells, and creating a raster map from the results. The raster map can contain real-world distances (*Appendix 1.12*) or raster distances (number of cells) from features or non-null cells in the input vector or raster map.
- **Neighborhood analysis**: Analyzing the neighborhood of the input raster. It usually involves convolution, which calculates some kind of statistics from the neighboring rasters of every cell, and writes the result in the appropriate cell of the output raster. The search radius can be circular or rectangular, and take an arbitrary size.

As you can see, the definitions have changed, as we cannot talk about geometries and attributes separately in case of raster data. Rasters offer full coverage of a rectangular area, therefore, if we use two perfectly aligned raster layers with coincident cells, the result will have the same cell number and cell size, and only the values matter. If not, a sophisticated GIS will resample one of the raster layers by simply aligning it with the other one, or interpolating its values during the process.

Multi-criteria evaluation

As we need to analyze the suitability of an area based on some preferences, we are basically doing an **MCDA** (**Multi-criteria decision analysis**). MCDA, in GIS, is generally done with raster data, and the final map shows the suitability of every cell in the study area. We can use MCDA for different purposes, like analyzing the suitability of the land for a specific species, or choosing the right site for a building with quantitative needs. During the process, we have to create raster maps for every criteria, then calculate the final suitability based on them. For this task, we differentiate between these two kinds of data:

- **Constraint**: Binary raster maps having cells with the value of zero (not suitable for the task), and having cells with the value of one (suitable for the task). These binary raster layers can be considered as masks, and define the areas we can classify in our final assessment.
- **Factor**: Raster maps showing the possibility that a cell will be suitable for a given criteria, also called fuzzy maps. Their values are floating point numbers between 0 and 1 (0 represents 0%--absolutely sure it is not suitable, while 1 represents 100% --absolutely sure the cell is suitable).

 Raw continuous data, such as distance from features, become fuzzy maps by using a normalization method. Don't worry about that at this point; we will discuss it later in this chapter (*Fuzzifying crisp data*).

In the end, we will have to create a single map by combining the different constraints and factors, showing the overall suitability of the cells calculated from the different factors, and masked by the union of the different constraints. There are several approaches and steps to execute an MCDA analysis, although in GIS, the most popular approach is to use the **multi-criteria evaluation** (**MCE**) method. By using this method alone, the result will have some uncertainty due to the involved subjectivity, although it will suit us in our task. First, let's break down our criteria to constraints and factors as follows:

- **Constraints**: Study area, maximum 5 kilometers from main roads, specific land use types, slope less than 10 degrees, minimum 200 meters away from waterways and water bodies
- **Factors**: Close to main roads, close to the mean point of the appropriate settlements, far from waterways and water bodies

Using this naive grouping, we have to process some of our data twice, as we have some overlaps between our constraints and our factors. However, we do not need to use those data as both constraints and factors. We can normalize our factors in a way that the constrained areas automatically get excluded from the result. Furthermore, as our DEM is already clipped to the borders of our study area, we do not have to create a raster layer from our study area. That is, we can regroup our tasks in the following way:

- **Constraints**: Specific land use types, slope less than 10 degrees
- **Factors**: Close to main roads (maximum 5 kilometers), close to the mean point of the appropriate settlements, far from waterways and water bodies (minimum 200 meters)

Creating the constraint mask

In order to create constraints, we need to convert our input features to raster maps. Before converting them, however, we need to open the correct layers, and apply filters on them to show only the suitable features:

1. Open the `landuse` layer and the SRTM DEM.

2. Apply a filter on the `landuse` layer to only show features which are restricted. It is simpler to create a filter which excludes land use types suitable for us, as we have fewer of them. Let's assume grass and farm types are suitable, as we can buy those lands. The only problem is that QGIS uses GDAL for converting between data types, which does not respect filtering done in QGIS. To overcome this problem, apply a filter on the layer with the expression `"fclass" != 'grass' AND "fclass" != 'farm'`, then save the filtered layer with **Save As**:

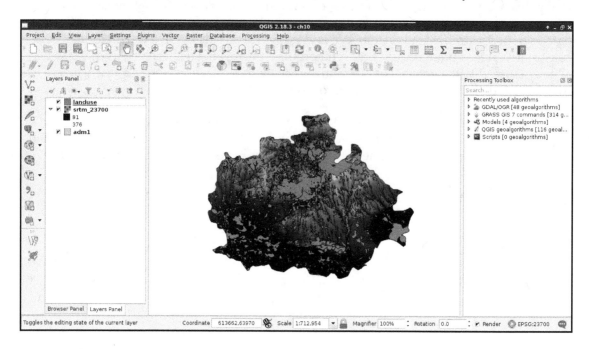

A more optimal way would be to select features from the `landuse` layer suitable for us. On the other hand, we would need a vector layer completely covering our study area for that. As our `landuse` layer has partial coverage, we select features not suitable, and invert the result later.

The next step is to create the required raster layers. This step involves calculating slope values from the DEM, and converting the vector layers to rasters:

1. Calculate the slope values using **Raster** | **Terrain Analysis** | **Slope** from the menu bar. The input layer is the DEM, while the output should be in our working folder. The other options should be left with their default values.

 The **Slope** tool outputs the slope values in degrees. However, other more sophisticated tools can create outputs with percentage values. If expressed as a percentage, a 100% slope equals to 45 degrees.

2. Right-click on the DEM, and select **Properties**. Navigate to the **Metadata** tab, and note down the resolution of the layer under the **Pixel Size** entry. We could use more detailed maps for our vector features, however, as the resolution of our coarsest layer defines the overall accuracy of our analysis, we can save some computing time this way.

3. Convert the filtered land use layer to raster with the **Raster** | **Conversion** | **Rasterize** tool. The input layer should be the filtered `landuse` layer, the output should be in our working folder, while the resolution should be defined with the **Raster resolution in map units per pixel** option with the values noted down before. The attribute value does not matter, however, we should use absolute values for the resolutions. The order of the values noted down matches the order we have to provide them (horizontal, vertical).

4. Define our project's CRS on the resulting raster layer to avoid confusion in the future (**Properties** | **General** | **Coordinate reference system**):

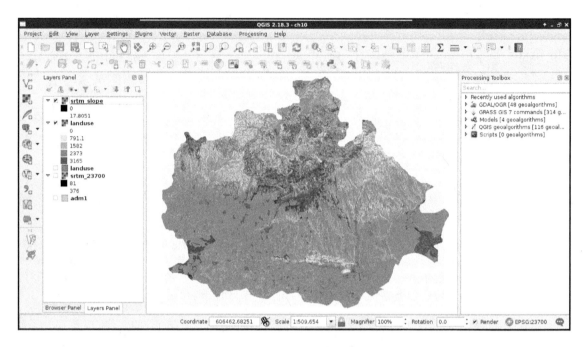

Now we have a problem. Our land use raster's extent is limited to the extent of the land use vector layer. That is, the raster does not cover our study area. If we leave it like this, we instantly fail one of our criteria, as we do not analyze the whole study area. We can overcome this issue by creating a base raster. The **Rasterize** tool has an option to overwrite an existing raster, and **burn** the rasterized features in it:

1. Create a constant raster with **QGIS geoalgorithms** | **Raster tools** | **Create constant raster layer**. The reference layer should be the slope layer, as it covers the whole study area. The constant value should be 0. We can overwrite our land use raster with the output of this file.

2. Use the **Rasterize** tool again. The input should be the land use vector layer, while the output should be the constant raster we overwrote our land use raster layer with. We should keep the existing size and resolution this time (default option).

Now we have a continuous and a discrete raster layer, which should create a mask together somehow. Using vector data, we can easily overlay two layers, as both consist of the same types--geometries. We can compare geometries safely, and get geometries as a result. However, in case of raster data, the **geometries** are regular grids, and overlaying them makes little sense for any analysis. In this case, we overlay cell values which represent some kind of attribute. Considering this, how can we compare two completely different values? What can be the result of overlaying slope degrees and land use IDs? What is the intersection of 15° and 2831? The answer is simple--we can only get meaningful overlays from comparable layers. That is why we need to convert our slopes and land use to constraints--0% suitability and 100% suitability values.

When we assign new values to raster layers based on some rules, it is called reclassification. We can reclassify raster layers in QGIS by using the raster calculator. Let's open it from **Raster | Raster Calculator**. The raster calculator in QGIS is somewhat similar to the field calculator, although it has limited capabilities, which include the following:

- **Variables**: Raster bands from raster layers. Only a single band can be processed at a time, although we have access to different bands of multiband rasters by referencing their band numbers (for example, `multiband@1`, `multiband@2`, `multiband@3`, and so on).
- **Constants:** Constant numbers we can use in our formulas.
- **Operators:** Simple arithmetic operators, power, and the logical operators `AND` and `OR`.
- **Functions:** Trigonometric and a few other mathematical functions.

- **Comparison operators:** Simple equality, inequality, and relational operators returning Booleans as numeric values. That is, if a comparison is `true`, the result is `1`, while if it is `false`, the result is `0`:

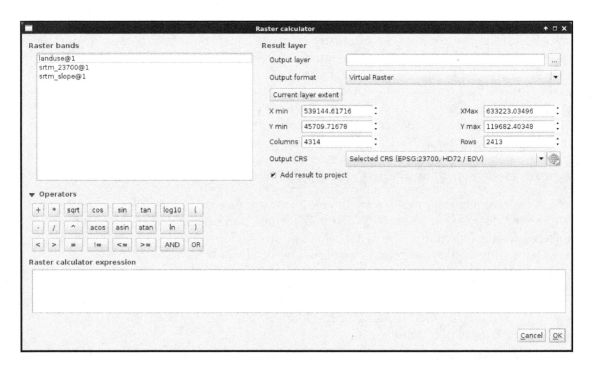

Always watch out for the current extent! You can load the extent of any processed raster layer by selecting it and clicking on **Current layer extent**. Make sure that you use the extent of the processed raster layer and not any other extent. Otherwise, QGIS may crop the layer, creating an incorrect result.

With these variables, constants, and operators, we need to create a function or expression which iterates through every cell of a single, or multiple raster layers. The resulting raster will contain the results of the function applied to the individual cells. As our constraint maps should only contain binary values, we can get our first results easily by using simple comparisons as follows:

1. Reclassify the land use raster using the raster calculator. The rule is, every raster with an ID greater than zero should have the value of 0 (not suitable), while cells with zero values should get a value of 1 (suitable). We can use an expression like `"landuse@1"` = 0. The output should be a `GeoTIFF` file saved in our working folder.

2. Reclassify the slope raster using the raster calculator. We need every cell containing a slope value less than 10° to get a value of 1. Other cells should get a value of 0. The correct expression for this is `"srtm_slope@1" < 10`. Similar to the previous constraint, the output should be a `GeoTIFF` in our working folder:

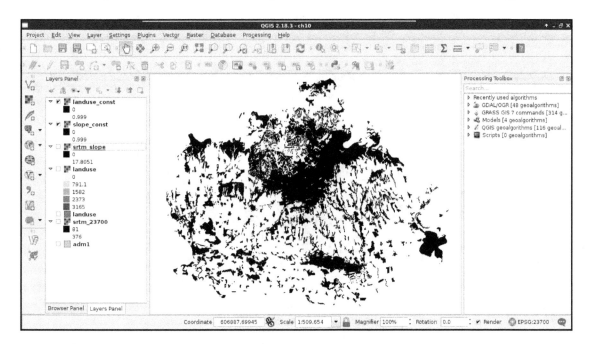

Don't worry about the maximum value of `0.999` in the **Layers Panel**. Remember, QGIS uses a cumulative cut when displaying raster layers, thus, cuts the top and bottom 2% of the values.

Now we have two binary constraint layers, which can be directly compared, as their values are on the same scale. Using binary layers A and B, we can define the two simplest set operations as follows:

- **Intersection (A × B)**: The product of the two layers results in ones where both of the layers have ones, and zeros everywhere else.
- **Union (A + B - A × B)**: By adding the two layers, we get ones where any of the layers has a one. Where both of them have ones (in their intersections), we get twos. To compensate for this, we subtract one from the intersecting cells.

What we basically need is the union of constraints (zeros). Logically thinking, we can get those by calculating the intersection of suitable cells (ones). Let's do that by opening the raster calculator, and creating a new GeoTIFF raster with the intersection of the two constraint layers as follows:

```
"landuse_const@1" * "slope_const@1"
```

Now we can see our binary layer containing our aggregated constraint areas:

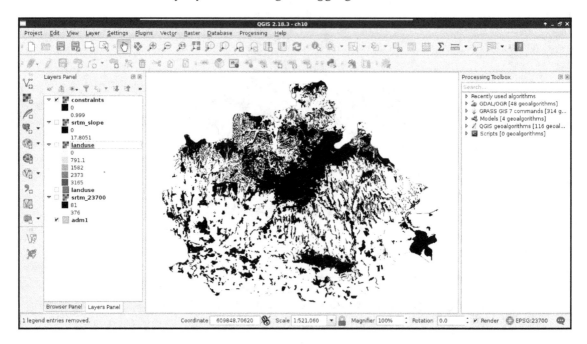

Using fuzzy techniques in GIS

Now that we have our final constraint layer, which can be used as a simple mask, we can proceed and create our factors. First, we can remove every intermediary layer we worked with, as our factors use different vector layers as input:

1. Open the `geonames`, `rivers`, `waterbodies`, and `roads` vector layers.
2. Filter the `geonames` layer to only show the seats of the administrative regions. The correct expression is `"featurecod" LIKE 'PPLA%'` or `"featurecode" LIKE 'PPLA%'` depending on which version we use.

3. Filter the `roads` layer to only show motorways and highways. Such a filter can be applied with the expression `"fclass" LIKE 'motorway%' OR "fclass" LIKE 'primary%'`.

4. Get the mean point of the seats of the filtered settlements by using the **QGIS geoalgorithms | Vector analysis tools | Mean coordinate(s)** tool. The input should be the filtered `geonames` layer, while the rest of the options can be left with their default values.

5. Save every result (that is, filtered `roads`, `mean_coordinates`, `waterways`, and `waterbodies`) to the working folder with **Save As**:

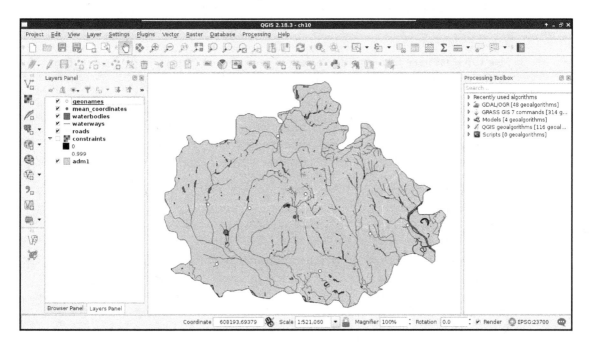

Proximity analysis with rasters

The easiest way to carry out a proximity analysis using rasters is GDAL's **Proximity** tool in QGIS. The tool requires a raster layer where features are described by cell values greater than zero. It takes the input raster, and creates the proximity grid--a raster with the same extent and resolution filled with distances from cells with values greater than zero.

The behavior of the **Proximity** tool implies the following two things:

- We need to rasterize our input features
- We need to supply our rasterized features in a raster map covering our study area

As we've already found out, we can supply an existing raster layer to the **Rasterize** tool:

1. Select one of the factor inputs (like `waterways`, `mean_coordinates`, and so on).
2. Create a constant raster (a raster, where every cell has a same value) with the tool **QGIS geolagorithms | Raster tools | Create constant raster layer**. Supply the value of 0, and the constraints layer as a reference. Save it using the name of the selected factor.
3. Use the **Rasterize** tool with the selected factor's vector layer and the constant raster map created in the previous step.

> If you cannot see the rasterized features in the resulting layer, you can use the actual minimum and maximum values in **Properties | Style | Load min/max values**. If you still cannot see anything, make sure to select the **Actual (slower)** option in the **Accuracy** menu, and load the values again.

4. Use the **Raster | Analysis | Proximity** tool to calculate the distances between zero and non-zero cells. The default distance units of **GEO** is sufficient, as it will assign values based on great-circle distances in meters. Save the result as a new file in a temporary folder.
5. Clip the result to the study area using **Raster | Extraction | Clipper**. Use the already extracted study area as a mask layer. Specify to cut the extent to the outline of the mask layer. Specify `-1` as **No data value**, as 0 represents valuable information for the analysis.
6. Remove the temporary layer.
7. Repeat the steps with every input factor.

> As GDAL warns you after finishing with a distance matrix, using non-square rasters reduces the accuracy of the analysis. The approximation that GDAL's **Proximity** tool creates is now enough for us. If you need more accurate results in the future, you can use GRASS's **r.grow.distance** tool from **GRASS GIS 7 commands | Raster**.

The distance matrices visualized in QGIS should have a peculiar texture slightly resembling a beehive:

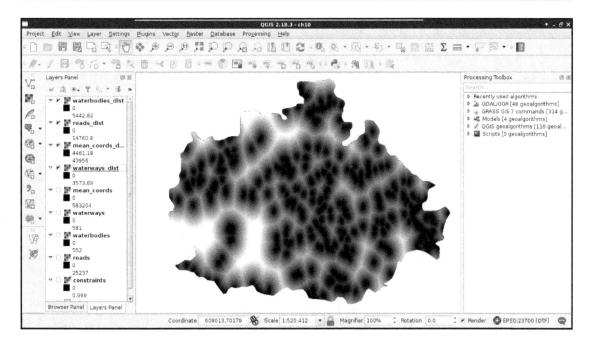

 Wondering if it would be easier to clip the layer we use as a basis for the rasterization? It would be if GDAL's **Proximity** tool didn't handle NoData values as features introducing implausible edge effects to our analysis (*Appendix 1.13*).

Now that we have the distance matrices we will use for our factors, we can get rid of the intermediary data (that is, vectors and rasterized features). The next problem is that we have a single criterion in two different layers. We need distances from waters, although we have distances from rivers and lakes separately. As both of them form the same preference, and their units are the same (that is, they are on the same scale), we can use set operations to make a single map out of them. The two essential set operations for non-binary raster layers **A** and **B** using the same scale look like the following:

- **Intersection (min(A, B))**: The minimum of the two values define their intersection. For example, if we have a value of 10% for earthquake risk and a value of 30% for flood risk, the intersection, that is the risk of floods and earthquakes is 10% (not at the same time, though--that is an entirely different concept).

- **Union (max(A, B))**: The maximum of the two values define their union. If we have the same values as in the previous example, the risk of floods or earthquakes is 30%.

For creating the final water distance map, we need the intersection of the `waterways` and `waterbodies` layers. Unfortunately, we do not have minimum and maximum operators in QGIS's raster calculator. On the other hand, with a little logic, we can get the same result. All we have to do is composite two expressions in a way that they form an if-else clause:

```
("waterbodies_dist@1" <= "waterways_dist@1") * "waterbodies_dist@1"
 + ("waterbodies_dist@1" > "waterways_dist@1") * "waterways_dist@1"
```

This preceding expression can be read as follows:

- If cell values from `waterbodies_dist` are equal or smaller than cell values from `weterways_dist`, return one, otherwise return zero. Multiply that return value with the `waterbodies_dist` layer's cell value.
- If cell values from `waterbodies_dist` is larger than cell values from `waterways_dist`, return one, otherwise return zero. Multiply that return value with the `waterways_dist` layer's cell value.
- Add the two values together:

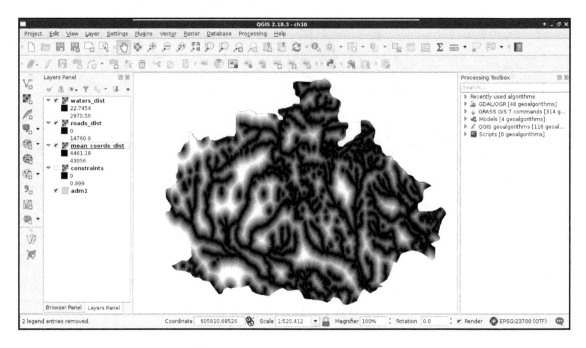

As we have the final distance layer for waters, the `waterways` and `waterbodies` layers are now obsolete, and we can safely remove them.

Fuzzifying crisp data

What we have now are three layers containing raw distance data. As these data are part of different criteria, we cannot directly compare them; we need to make them comparable first. We can do this by normalizing our data, which is also called fuzzification. Fuzzy values (μ) are unitless measures between 0 and 1, showing some kind of preference. In our case, they show suitability of the cells for a single criterion. As we discussed earlier, 0 means 0% (not suitable), while 1 means 100% (completely suitable).

The problem is that we need to model how values between the two edge cases compare to the normalized fuzzy values. For this, we can use a fuzzy membership function, which describes the relationship between raw data (crisp values) and fuzzy values. There are many membership functions with different parameters. The most simple one is the linear function, where we simply transform our data to a new range. This transformation only needs two parameters--a minimum and a maximum value. Using these values, we can transform our data to the range between **0** and **1**. Of course, a linear function does not always fit a given phenomenon. In those cases, we can choose from other functions, among which the most popular in GIS are the sigmoid and the J-shaped functions:

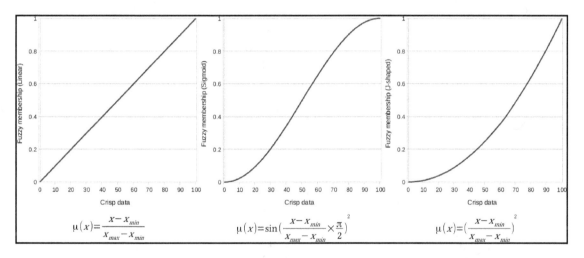

$$\mu(x) = \frac{x - x_{min}}{x_{max} - x_{min}}$$

$$\mu(x) = \sin\left(\frac{x - x_{min}}{x_{max} - x_{min}} \times \frac{\pi}{2}\right)^2$$

$$\mu(x) = \left(\frac{x - x_{min}}{x_{max} - x_{min}}\right)^2$$

There are various other formulae for fuzzifying crisp data (*Appendix 1.14*), however, most of them use more parameters, therefore, need more considerations. You can read more about these simple transformations at GRASS GIS's r.fuzzy addon's manual page at `https://grass.osgeo.org /grass64/manuals/addons/r.fuzzy.html`.

To solve this problem, we have to interpret the membership functions, and choose the appropriate one for our crisp data. The various membership functions are explained as follows:

- **Linear**: The simplest function, which assumes a direct, linear relationship between crisp and fuzzy values. It is good for the `roads` layer with the minimum value of `0` and the maximum value of `5000`. We have to handle distances over 5000 meters manually, and invert the function, as cells closer to the roads are more suitable.
- **Sigmoid**: This function starts slowly, then increases steeply, and then ends slowly. It is good for the mean coordinates map, as the benefit of being near to the settlements' center of mass diminishes quickly on the scale of the whole study area. We have to use the minimum value of `0`, and the maximum value of the layer. Additionally, we have to invert the function for this layer, too.
- **J-shaped**: A quadratic function which starts slowly, and then increases rapidly. It can be used with the waters layer, as it is safer to assume a quadratic relationship on a risk factor, when we do not have information about the actual trends. We can use the minimum value of `200` and the maximum value of the layer.

You can easily invert a fuzzy membership function by subtracting the values from one, as fuzzy values are between `0` and `1`. If you use this method on a fuzzy layer, you can get its complement layer.

First, let's use the J-shaped membership function on the `waters` layer, as follows:

1. We have a minimum value of 200 meters at hand, however, we need to find out the maximum value. To do this, go to **Properties | Metadata**. The `STATISTICS_MAXIMUM` entry holds the maximum value of the layer. Round it up to the nearest integer, and note down that number.
2. Open a raster calculator, and create an expression from the J-shaped function, the minimum, and the maximum values. Handle values less than the minimum value in a conditional manner. Save the result as a `GeoTIFF` file. The final expression should be similar to the following:

```
("waters_dist@1" <= 200) * 0 + ("waters_dist@1" > 200) *
(("waters_dist@1" - 200) / (max - 200)) ^ 2
```

Next, we should apply the linear membership function to the `roads` layer as follows:

1. Open a raster calculator, and create an expression from the inverted linear function, the minimum value of `0`, and the maximum value of `5000`. Handle values more than 5000 meters in a conditional manner. Save the result as a `GeoTIFF` file. The final expression should be similar to the following:

```
("roads_dist@1" > 5000) * 0 + ("roads_dist@1" <= 5000) *
(1 - ("roads_dist@1" / 5000))
```

By running the expression, we should be able to see two of our factor layers:

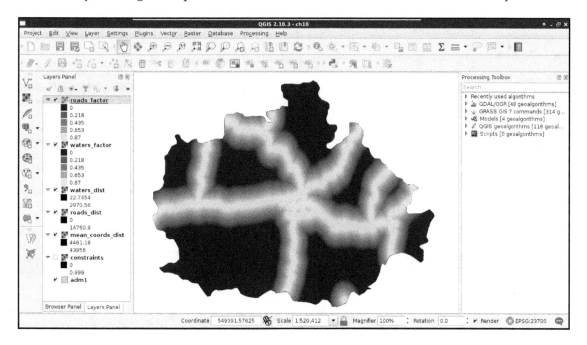

Finally, we use the sigmoid function for the mean coordinates layer like this:

2. We know the minimum value is `0`, however, we need to find out the maximum value. Check it in **Properties** | **Metadata**, round it up to the nearest integer, and note down the value.

3. Open a raster calculator, and create an expression from the minimum and maximum values, and from the inverted sigmoid function. We do not have access to π, although we can hard code it as `3.14159`. Save the result as a `GeoTIFF` file:

```
1 - sin("mean_coords_dist@1" / max * (3.14159 / 2)) ^ 2
```

Aggregating the results

Now that we have all of our factors set up, we only need to create a single factor map out of them. We could simply overlay them, and calculate their intersection for a restrictive, or their union for a permissive suitability. However, we can also consider them as components of a composite map, and calculate their average as follows:

1. Open a raster calculator, and calculate the factors' average by adding them together, and dividing the result by the number of components. Save the result as a GeoTIFF file:

```
("mean_coords_factor@1" + "roads_factor@1" +
 "waters_factor@1") / 3
```

The result should be a beautiful, continuous raster map showing our factors:

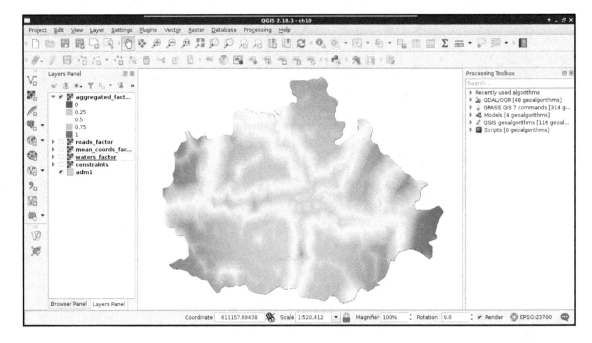

What is the problem with this approach? Some of our factors also contain constraints as zero values. By involving them in the calculation, the other two factors with higher suitability values can compensate those constrained cells. To get correct results, we have to manually handle the zero values in the roads and waters factor layers.

2. Open a raster calculator, and overwrite the aggregated factors with an expression handling zero values in the roads and waters factor layers. Make sure to save it as a `GeoTIFF` file.

```
("roads_factor@1" > 0 AND "waters_factor@1" > 0) *
 ("mean_coords_factor@1" + "roads_factor@1" +
  "waters_factor@1") / 3
```

By calculating the average of the factors, we assume their weights are equal in the analysis. This is not always true. Of course, you can think up weights, introducing another level of subjectivity into the analysis, but you can also try to calculate weights by defining the relative importance of the factors, comparing two of them at a time. This method is called **AHP (Analytic Hierarchy Process)**. There is a nice example on Wikipedia about this method at `https://en.wikipedia.org/wiki/Analytic_hierarchy_process_%E2%80%93_car_example`. There is also a great online AHP calculator at `http://bpmsg.com/academic/ahp_calc.php`.

Now we have a less beautiful, but correct result. The only thing left to do is to simply overlay the constraints map with the aggregated factors, which is done using the following expression:

```
"aggregated_factors@1" * "constraints@1"
```

When the raster calculator is finished, we should be able to see our final suitability map:

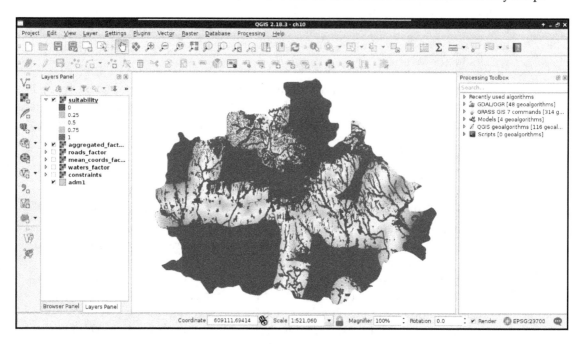

The final step is to defuzzify the final map to get crisp data that we can evaluate. This step is very simple in GIS, as the MCE is usually done in such a way that we can get percentage values if we multiply the fuzzy values by 100. As this is a very trivial operation, we do not even have to calculate the percentage map, only label the actual intervals:

Value	Label	Suitability
0-0.25	0-25%	Poor
0.25-0.5	25-50%	Weak
0.5-0.75	50-75%	Moderate
0.75-1	75-100%	Excellent

Calculating statistics

In GIS, statistics can be computed from both raster and vector data. However, even calculating raster statistics often involves some kind of vector data. For example, we would like to include some statistical indices in our assessment regarding the suitable areas. More precisely, we would like to include at least the minimum, maximum, and average slope, the minimum, maximum, and average suitability, the average distance from the mass point of the settlements, and the minimum distance from waters. For this task, we cannot use our rasters alone; we need to calculate indices from them only where they overlap with our suitable areas. For this, we need our suitable areas as polygons, and then we can leave the rest of the work to QGIS.

In order to get our suitable areas as polygons, we need to delimit them on our suitability layer. The most trivial first choice is to select every cell with an excellent rating. However, how many cells do we have with more than 75% suitability? If we have only a few, vectorizing them would make no sense, as every resulting polygon would fail the minimum area criterion. Furthermore, if we have some sites meeting the 1.5 km^2 criterion, but the main roads go right through them, that is also a failure, as we cannot have a single site divided by a high traffic road.

In order to get the minimum suitability value that our analysis is viable with, we can limit the suitability layer to a range. Let's open **Properties** | **Style**, and choose **Singleband gray** for **Render type**. Now we can manually input the range we would like to check (0.75 as **Min** and 1 as **Max** first), and set **Contrast enhancement** to **Clip to MinMax**. This way, QGIS simply does not render cells outside of the provided range. By using this representation model, we only have to load the roads layer, and measure some of the visualized patches. We can measure an area with the **Measure Area** tool from the main toolbar. We have to select it manually by clicking on the arrow next to **Measure Line**, and choosing it.

The tool works like the regular polygon drawing tool--a left-click adds a new vertex, while a right-click closes the polygon:

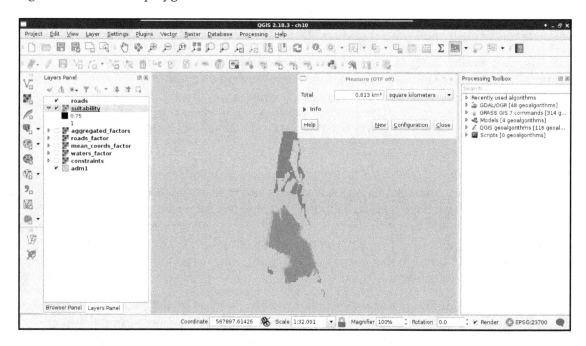

If you have very small patches, choose a lower **Min** value and repeat the process. Choose a minimum value where you have several suitable areas. For me, the value of `0.6` worked well, although it may change with the study area.

Vectorizing suitable areas

Now that we have an appropriate suitability value, we can vectorize our suitability map. We've already seen how vector-raster conversion works, but we did not encounter raster-vector conversion. As every raster layer consists of cells with fixed width and height values, the simplest approach is to convert every cell to a polygon. GDAL uses this approach, but in a more sophisticated way. It automatically dissolves neighboring cells with the same value. In order to harness this capability, we should provide a binary layer with zeros representing non-suitable cells, and ones representing suitable cells:

1. Open a raster calculator, and create a binary layer with a conditional expression using the minimum suitability value determined previously. Such an expression is `"suitability@1" >= 0.6`. Save the result as a `GeoTIFF` file.

2. Open **Raster** | **Conversion** | **Polygonize** from the menu bar.

3. Provide the binary suitability layer as an input, and specify an output for the polygon layer:

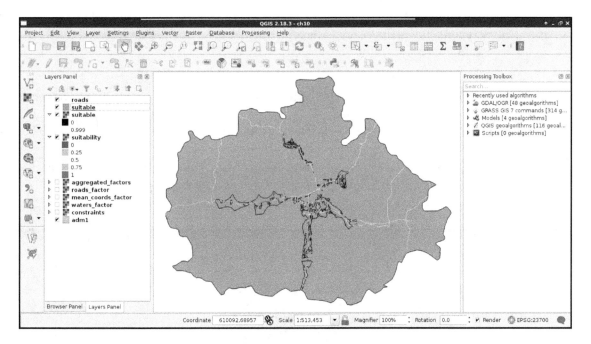

Now we have a nicely dissolved polygon layer with **DN** (**digital number**) values representing suitability in a binary format. We can apply a filter on the layer to only show suitable areas:

```
"DN" = 1
```

As the polygons do not respect the main roads, we need to cut them where the roads intersect them. This seems to be a trivial problem, although there are no simple ways to achieve this in QGIS. On the other hand, we can come up with a workaround, and convert our filtered polygons to lines, merge them with the roads, and create polygons from the merged layer.

4. Convert the filtered suitable areas layer to lines with **QGIS geoalgorithms** | **Vector geometry tools** | **Polygons to lines**. The output should be saved on the disk, as the merge tool does not like memory layers.

5. Merge the polygon boundaries with the roads layer by using **QGIS geoalgorithms** | **Vector general tools** | **Merge vector layers**. The output can be a memory layer this time.

6. Create polygons from the merged layer with **QGIS geoalgorithms** | **Vector geometry tools** | **Polygonize**. Leave every parameter with their default values, and save the result as a memory layer.

Be sure to use the **Polygonize** tool. There is another tool called **Lines to polygons**, however, it converts linestring features to polygons directly, creating wrong results.

7. Now we have our polygon layer split by the roads, however, we've also got some excess polygons we don't need. To get rid of them, clip the result to the original suitable areas layer with **QGIS geoalgorithms** | **Vector overlay tools** | **Clip**. Save the result as a memory layer.

8. Closely inspect the clipped polygons. If they are correctly split at the roads, and do not contain excess areas, we can overwrite our original suitable areas layer with this:

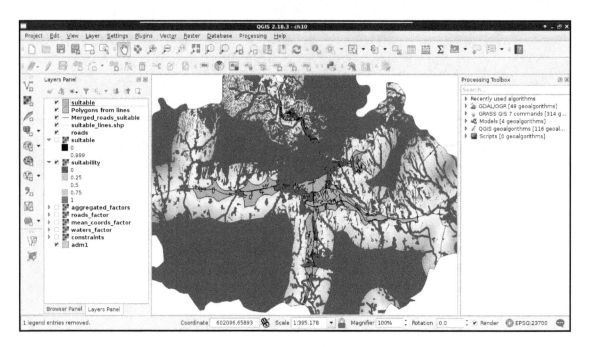

Don't worry if you get an error message saying QGIS couldn't save every feature because of a type mismatch. The clipped areas are stored in a polygon layer, therefore, the output layer's type will automatically be polygon. If QGIS detects that there are also some other types of geometries present in the saved layer, it still saves every matching feature. It just won't load the result automatically.

The last thing to do with our vector layer before calculating statistics is to get its attribute table in shape. If you looked at the attribute table of the polygonized lines, you would see that the algorithm automatically created two columns for the areas and the perimeters of the geometries. While we do not care about the perimeters in the analysis, creating an area column is very convenient, as we need to filter our polygons based on their areas. The only problem is that by clipping the layer, we unintentionally corrupted the area column. The other attribute we should add to our polygons is a unique ID to make them referable later:

1. Select the saved suitable areas polygon in the **Layers Panel**, and open a field calculator.
2. Check in the **Update existing field** box, and select the area column from the drop-down menu.
3. Supply the area variable of the geometries as an expression--$area, and recalculate the column.
4. Open the field calculator again, and add an integer field named id. The expression should return a unique integer for every feature, which is impossible to do in the field calculator. Luckily, we can access a variable storing the row number of every feature in the attribute table, which we can provide as an expression--$rownum.
5. Save the edits, and exit the edit session.

6. Apply a filter to only show the considerable areas using the expression `"area"` `>= 1500000`:

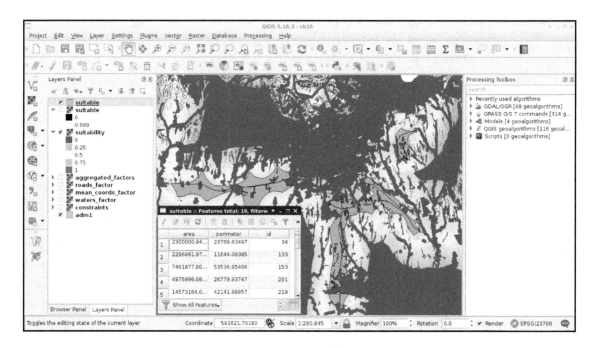

Using zonal statistics

Although calculating statistics from a whole raster layer has its own advantages, now we need raster statistics from only the portions overlapping with our suitable areas. We can do this kind of calculation automatically by using zonal statistics. Zonal statistics require a raster layer and a polygon layer as inputs, then creates and fills up attribute columns with all kinds of statistical indices (like count, sum, average, standard deviation, and so on) in the output polygon layer. In order to calculate all the required statistics, we need all the input raster layers first:

1. Open every raster layer needed for the statistics--the water distance, the mean coordinate distance, the slope, and the suitability layers.

2. Open the **Raster** | **Zonal statistics** | **Zonal statistics** tool.

3. Choose an appropriate raster layer as **Raster layer**, the suitable areas layer as the **Polygon layer containing the zones**, and supply a short prefix describing the raster layer (for example, mc_ for mean coordinates). Save the result as a memory layer. Check the appropriate indices, and uncheck the rest of them. Remember, water distance--minimum; mean coordinates--average (mean); slope--minimum, average, maximum; suitability--minimum, average, maximum.

4. Repeat the process for every input raster layer:

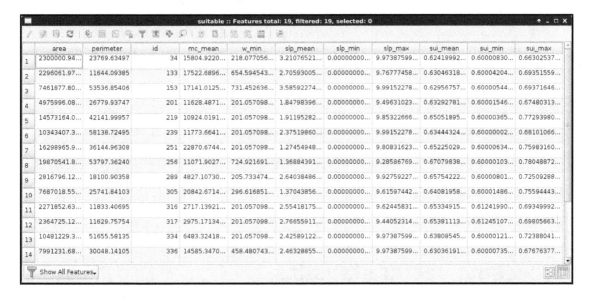

That's all. With a few clicks, we can get a lot of statistical indices from different raster layers and some polygons. On the other hand, those numbers are not comprehensive at all. For example, we do not know about the distribution of suitability values from some indices. As a matter of fact, having a histogram of the suitability values could enhance decision making, as we would see how common less suitable values, and the more suitable values in a site are. For that, we would need the histogram of the raster layer under our potentially suitable areas. Unfortunately, creating zonal histograms is not available in QGIS. Furthermore, the easiest approach involves a lot of manual labor. Let's create one or two histograms just to get the hang of it:

1. Open the attribute table of the suitable areas, and select the first row by clicking on the row number on the left.

2. Save the selected feature using **Save As**, and specifying **Save only selected features**.

3. Use the **Clipper** tool to clip the suitability raster layer to the saved feature.

4. Copy the style of the suitability layer, and paste it on the clipped suitability layer (this way, we get a colored line in the histogram).
5. Open **Properties** | **Histogram** on the clipped raster, and save the histogram as a PNG file with the **Save plot** button. Use the ID of the selected feature in the file name (for example, `histo_34.png`).

> You can speed up this manual process somewhat by using the **QGIS geoalgorithms** | **Vector general tools** | **Split vector layer** tool with the `id` column of the suitable areas. It saves features with the same IDs on different layers in the output folder. Then you can use **GDAL/OGR** | **[GDAL] Extraction** | **Clip raster by mask layer** as a batch process (right-click on it, and select **Execute as batch process**) to create every extraction at once. You still have to save the histograms manually, though.

There are still several problems with this approach, although this is the closest we can get to a histogram in QGIS without scripting in Python or R. The problems include the following:

- The values are not binned. We have every different value as a single interval, making the histogram noisy.
- The frequency is expressed in cell counts. It would be much more clear if the frequency would be expressed in percentage values.

Accessing vector statistics

Getting vector statistics in QGIS is very straightforward. The method is similar to raster statistics, although as we can store as many attributes as we want in a vector layer, we can only calculate statistics from a single numeric column at a time. We can access the **Show statistical summary** tool from the main toolbar (purple Σ button), choose a layer, then choose a numeric column. To save the statistics to a file, we can use **QGIS geoalgorithms** | **Vector table tools** | **Basic statistics for numeric fields**. We can also calculate grouped statistics with **QGIS geoalgorithms** | **Vector table tools** | **Statistics by categories**.

Creating an atlas

The atlas generator is the most powerful feature of QGIS's print composer. It can create a lot of maps automatically based on a template we provide. The underlying concept is very basic--we have to provide a polygon layer with a column which has unique values.

The print composer takes that layer, and creates a separate map page for every different value (therefore, feature) it can find in the provided column. Furthermore, it grants access to the current feature it uses for the given page. The real power comes from the QGIS expression builder, which enables us to set cartographic preferences automatically. With this, we can build a template for our atlas and use it to showcase each suitable area in its own map.

First of all, if we would like to create a front page with every suitable area on a single map, we have to create a feature enveloping the polygons from our suitable areas polygon layer. We can create such a polygon with **QGIS geoalgorithms** | **Vector geometry tools** | **Convex hull**. It takes a vector layer as an argument, and creates a single polygon containing the geometries of the input features:

1. Create the convex hull of the suitable areas using the aforementioned tool. Save the output to the working folder, as the merge tool does not like memory layers.
2. Open the attribute table of the convex hull layer. Remove every attribute column other than id. They would just make the merged layer messier, as the merge tool keeps every attribute column from every input layer. Don't forget to save the edits, and exit the edit session once you've finished.
3. Merge the suitable areas layer and the convex hull layer with **Merge vector layers**. Save the output in the working folder with a name like coverage.shp.

Now we have every page of our atlas in the form of features in our coverage layer. We can proceed and make it by opening **New Print Composer**, and using the **Add new map** tool to draw the main data frame. We should leave some space for the required information on one of the sides. In order to begin working with an atlas, we have to set some parameters first:

1. Go to the **Atlas generation** tab on the right panel.
2. Provide the coverage layer as Coverage layer, and the id column as Page name. Check the **Sort by** box, and select the id field there too.
3. Select the map item, navigate to the **Item properties** tab, and check the **Controlled by atlas** box. This is an extension to the extent parameters, which automatically sets the data frame's extent to the extent of the current feature. Select the **Margin around feature** option.

4. Click on **Preview Atlas** on the main toolbar. You should be able to see the first page instantly, and navigate between the different pages with the blue arrows:

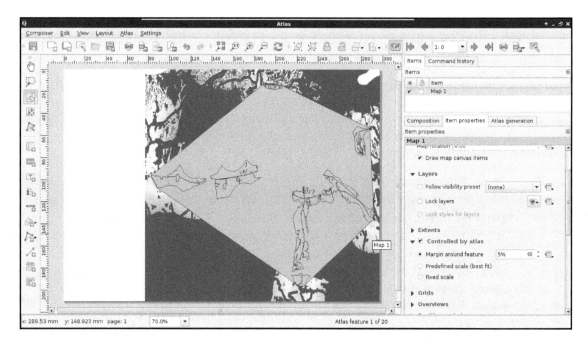

As the next step, we should style our layers in a way that they create an aesthetic composition in our atlas. For example, we should make the convex hull invisible, and remove the fills from the suitable sites:

1. Open the **Properties** | **Style** menu of the suitable sites layer, and select **Rule-based** styling.
2. Modify the existing rule. Name it `selected`, and create an expression to show the current atlas feature if it is not the convex hull. Such an expression is `"id" > 0 AND @atlas_featureid = $id`, hence, the convex hull has an ID of `0`. Style it with only an outline (**Outline: Simple line**), and apply a wide, colored line style.

You can reach every atlas-related variable in the **Variables** entry of the expression builder.

3. Add a new rule. Name it `not selected`, and create an expression to show every feature besides the current atlas feature and the convex hull. The correct expression is `"id" > 0 AND @atlas_featureid != $id`. Style them with a narrow black outline.

4. The dominance of zero values in the suitability layer distorts the look of the map. Classify zeros as null values by opening **Properties** | **Transparency**, and defining `0` in the **Additional no data value** field:

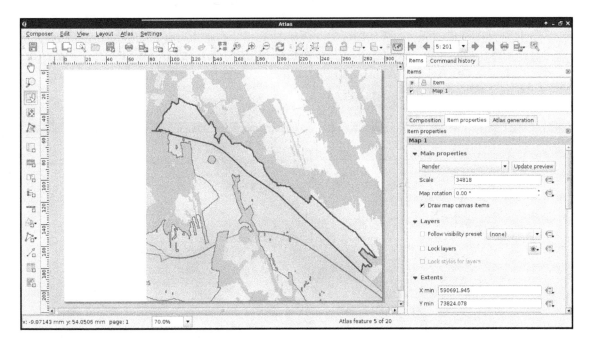

You can add a nice touch to your map by using something like OpenStreetMap as a base layer (*Appendix 1.15*). All you have to do is install **OpenLayers Plugin** (**Plugins** | **Manage and Install Plugins**), and select **OpenLayers plugin** | **OpenStreetMap** | **OpenStreetMap** from the new **Web** menu. Note that this procedure sets the projection to **EPSG:3857** automatically.

The second item we should add to our atlas is an overview map. This way, we can make sure we know where we are in the study area every time:

1. Add a new map frame with **Add new map** in one of the free corners of the canvas.
2. Style the layers in QGIS as you see fit. For the sake of simplicity, I added only the study area's polygon and the water layers.
3. After styling, go back to the composer, select the overview map, and check the **Lock layers box.**
4. Position the map with **Move item content** in a way that the whole study area is in the frame. You can use the **View extent in map canvas** button as initial guidance.
5. In the **Overviews** section add a new item. Select the other map as **Map frame**.
6. Restore the initial layers in QGIS:

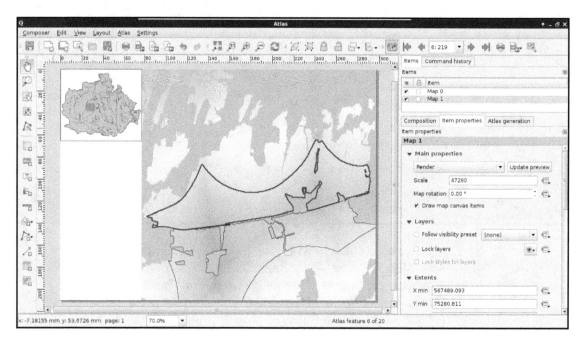

The next item we add is one of the most important parts of our atlas. It is the attributes of the atlas features. A simple way to achieve this would be to add an attribute table item with **Add attribute table**, although it cannot be customized enough to fit in our atlas. For these cases, QGIS's print composer offers a highly customizable item--the HTML frame. With that item, we can visualize any valid HTML document. Furthermore, we can use the expression builder to write expressions, which will be evaluated by QGIS and rendered in the HTML frame:

1. Add a new HTML frame with the **Add HTML frame** tool on the left toolbar.
2. In its **Item properties** dialog, select the **Source** radio button, and **Evaluate QGIS expressions in HTML source** box.

Now we just have to write our HTML containing the attributes of the features. A great thing in HTML is that plain text is a completely valid element. Therefore, we only need to know how to use one HTML element in order to fill our HTML frame, which is as follows:

- **
**: Inserts a line break in the HTML source.

The rest of the task is simple string concatenation (| | operator). We evaluate the attributes of the features (ID, area, and statistics), then concatenate them with the rest of the text and the <br> elements. Furthermore, as the HTML frame is an atlas-friendly item, the attributes of the current feature are automatically loaded, therefore, we can refer to the correct attribute with the name of the column. Finally, as the statistical indices are quite long, we should round them off with the round function. We can also divide the area by 1000000 to get the values in km^2:

```
[%'ID: ' || "id" || '<br>' ||
'Area: ' || round(("area" / 1000000), 1) || ' km²<br>' ||
'Min. suitability: ' || round("sui_min", 1) || '<br>' ||
'Avg. suitability: ' || round("sui_mean", 1) || '<br>' ||
'Max. suitability: ' || round("sui_max", 1) || '<br>' ||
'Min. slope: ' || round("slp_min", 1) || '°<br>' ||
'Avg. slope: ' || round("slp_mean", 1) || '°<br>' ||
'Max. slope: ' || round("slp_max", 1) || '°<br>' ||
'Min. distance to waters: ' || round("w_min", 1) || ' m<br>' ||
'Avg. distance to center of mass:
' || round(("mc_mean" / 1000), 1) || ' km'%]
```

Before clicking on the **Refresh HTML** button, copy the content of the HTML source. If QGIS drops the expression, paste back the copied source, and click on **Refresh HTML** again.

We should expand our expression a little bit. Although it shows the attributes of the atlas features nicely, we get a bunch of irrelevant numbers on the first page. Instead of visualizing them, we should print the title of the project, and the attributions on the first page. We can easily do this by extending our expression with the CASE conditional operator. We just have to specify the ID of the convex hull in the CASE clause, and put the attributes of the atlas features in the ELSE clause:

```
[%CASE WHEN "id" = 0 THEN
'Assessment of suitable areas for building a ' ||
```

```
'warehouse in Baranya county. <br> Created ' ||
'by: G. Farkas 2017 <br> OpenStreetMap data ' ||
'© OpenStreetMap Contributors'
ELSE
'ID: ' || "id" || '<br>' ||
'Area: ' || round(("area" / 1000000), 1) || ' km²<br>' ||
'Min. suitability: ' || round("sui_min", 1) || '<br>' ||
'Avg. suitability: ' || round("sui_mean", 1) || '<br>' ||
'Max. suitability: ' || round("sui_max", 1) || '<br>' ||
'Min. slope: ' || round("slp_min", 1) || '°<br>' ||
'Avg. slope: ' || round("slp_mean", 1) || '°<br>' ||
'Max. slope: ' || round("slp_max", 1) || '°<br>' ||
'Min. distance to waters: ' || round("w_min", 1) || ' m<br>' ||
'Avg. distance to center of mass: ' || round(("mc_mean" / 1000),
1) || ' km'
END%]
```

Now we can see our attributes when we focus on a feature, while the overview page shows only attribution:

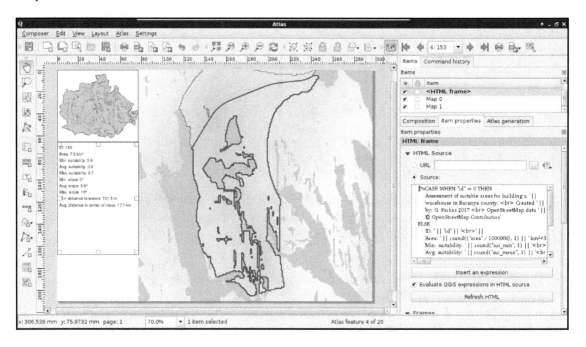

You can use any valid HTML syntax in the source. The only drawback of this item is that you cannot alter the style of the content directly from QGIS. You have to write CSS code for this in the **User stylesheet** field after activating it.

The final item we should add to our atlas is an image showing histograms. With data-defined override, we can get the correct histogram of every atlas feature if they are saved in the same folder, and contain the IDs of the features in their names:

1. Add a new image item with the **Add image** tool.
2. Choose the **Data defined override** button next to the `Image source` field, and select the **Edit** option.
3. Create an expression which returns the correct histogram file for every atlas feature. My expression is `'/home/debian/practical_gis/results/mce/histo/histo_' || attribute(@atlas_feature, 'id') || '.png'`. Note that the current atlas feature is not evaluated in a data-defined override (that is, we cannot access its attributes directly).

When we see our composition, we will be able to see some histograms, if we have them saved as images:

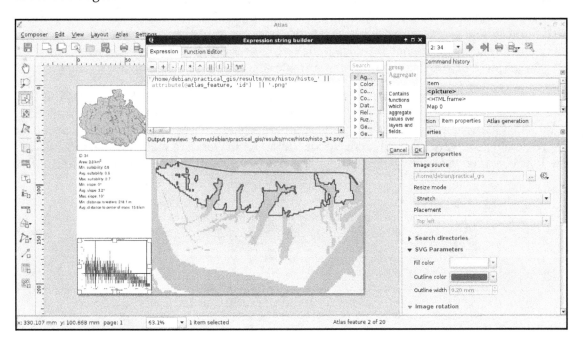

 You can include your company's logo on the first page easily by saving it along with the histogram images with a name reflecting the convex hull's ID.

The only thing left to do is to export our atlas. By selecting the **Export Atlas as Images** button on the main toolbar, we can see that the atlas can be exported in image, SVG, and PDF formats. The most convenient way of exporting is to save the entire atlas in a single PDF, where every atlas page is rendered on a separate page:

1. Select the **Atlas generation** tab in the right panel.
2. Check the **Single file export when possible** box.
3. Save the atlas as a PDF with **Export Atlas as Images | Export Atlas as PDF**.

Summary

Congratulations! You just carried out your first raster analysis with MCE. It is valuable to have the ability to use raster data effectively, and to create suitability analysis on demand. The best thing is that we were able to do the analysis with free and open source data. In this chapter, we learned how to effectively use and analyze raster data. We created a suitability map for an imaginary cause of building a warehouse based on a set of criteria. We were able to delimit suitable areas, and calculate statistics on them to further help making a good decision. Finally, we automated the map making process by creating an atlas with every important piece of information on every suitable site.

In the next chapter, we will leave the realm of desktop GIS, and dwell on web mapping. We will learn the basics of the server side of web mapping systems. First, we will serve our data with QGIS for instant and easy publication, then start to learn GeoServer, which is a far more capable software for publishing spatial data on the web. We will learn the most basic standards we can use with web mapping, and how we can utilize them for different results.

11
Showcasing Your Data

In the previous chapter, we learned about advanced raster analysis. We learned how to use the most essential raster-based tools by doing a complete suitability analysis based on open data. In order to make a better decision, we calculated some statistical indices on our suitable areas. Finally, to complete the assignment, we created an atlas from the results.

In this chapter, we will learn the basics of web mapping, more precisely, how client-server interaction works in the realm of spatial information technologies. We will focus on the server side, and on one of the most important aspects of web mapping--spatial data exchange. We will learn about spatial data formats and services readable by browsers, therefore, widely used for web mapping. Finally, we will learn how to use QGIS Server and GeoServer for publishing data on the web.

In this chapter, we will cover the following topics:

- Open Web Services
- Spatial data formats in web mapping
- Publishing spatial data with QGIS Server
- Publishing spatial data with GeoServer

Spatial data on the web

In order to understand how spatial data on the web works, we first need to get a picture about the architecture of the web. The web resides on the Internet, where we have to deal with two kinds of software--servers and clients:

- **Server**: An application that listens to a port with a background process (daemon), and accepts requests from that port. It processes valid requests and serves data according to them.

- **Client**: An application responsible for sending valid requests to server side application(s) (for example, web browser-web server, SSH client-SSH server, QGIS-PostgreSQL). It also needs to be able to interpret the response sent back from the server.

> If you are familiar with the client-server architecture, you can skip the following subsection.

Understanding the basics of the web

The Internet is designed to be an infinitely expandable network of computers. Therefore, servers and clients are only the end points of this network--there are additional nodes doing other tasks. For example, there are DNS servers, which map IP addresses to domain names; routers and switches forward the traffic. The web is one of the largest portions of the Internet, sharing specific, standardized content between end points. For creating a web application, the midpoints are out of concern. We only need to know how to configure web servers (backend), and how to write content for web clients (frontend).

> There are a lot of other use cases of the Internet. Just think of video streaming, direct file sharing (FTP), remote administration (SSH), or playing video games online.

In order to have a working architecture, the web is powered by standards instead of software. These are open standards, which define the intended behavior of every step in serving and receiving data on the web, mostly maintained by a large number of experts and companies forming the **World Wide Web Consortium** (**W3C**). This way, anyone can develop a web server or a web client, which is guaranteed to work with any website if these standards are followed. If not, for example, a web browser that places two line breaks on every <br> element, it is called a bug. No matter if the developers reason that **this is an intended feature, as two-line breaks look much better than a single one, the standard is wrong**. What do we gain from this strong standardization? We do not have to worry about compatibility issues. The standards make sure we can use Apache, nginx, Node.JS, or any other web server application as a web server, and the hosted files will work on any web browser following them. We only need to make sure that the web server we choose is capable enough for our needs, and the configuration is correct.

These standards are very specific, therefore, very long and complex. That is why we won't discuss them in detail but grab some of the more important parts from our perspective. In a web architecture, we have a web server, and a client capable of communicating with it (most often a web browser):

- **Web server**: A server application capable of communicating over the **hypertext transfer protocol** (**HTTP** or **HTTPS**). By default, web servers using HTTP listen on port 80, while the ones using HTTPS listen on port 443. The main responsibility of web servers is to accept HTTP requests, resolve paths, and serve content accordingly. Different web servers have different capabilities, although encrypting data (HTTPS) and compressing responses are often implemented. Web servers can access a portion of the server machine's file system from where they can serve these two kinds of resources:
  - **Static files**: HTML, CSS, JS files, images, and other static resources for the served web page.
  - **CGI**: Server-side scripts that the web server can call with parameters defined in the request as arguments. It resembles a command-line call with the difference that CGI programs must conform to web standards. CGI scripts can be written in any language the server's OS can run as a command-line program (most often, PHP, Python, Ruby, or C).

- **Web browser**: A client application capable of communicating over the hypertext transfer protocol (HTTP or HTTPS). It can send requests to web servers, and interpret responses. It can handle various types of data like the following received from a web server:
  - **Plain text**: The most basic response type. The browser renders it as plain text.
  - **Structured text**: Markup languages (like HTML and XML), CSS stylesheets, JS programs. The browser parses them, then creates a **Document Object Model** (**DOM**), preserving the structure and hierarchy of the source documents. It styles the DOM elements according to the rules in the stylesheets, and interprets the content of the JS files, allowing the JS programs to run on the client.
  - **Media elements**: **RGB** or **RGBA** (**red**, **green**, **blue**, **alpha**) images in common formats (like PNG, JPEG, BMP, and GIF), video files (WEBM, OGG, and MP4), subtitles, audio files (OGG and MP3). The client incorporates these elements into its DOM structure, rendering them in a usable way.

We can see the generalized scheme of the client-server architecture in the following figure:

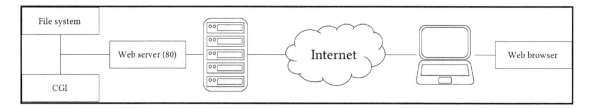

The next step is the communication between web servers and web clients. There are various standardized requests that a client can send to a web server, which serves content accordingly. The response is also standardized, therefore, the client can interpret it:

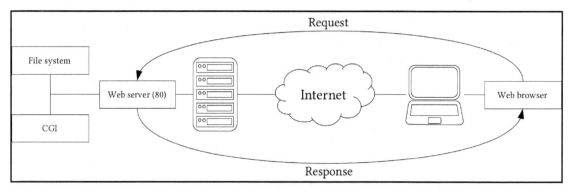

- **Request**: The web client sends a request to a destination identified with a URL. The URL contains the destination server machine's IP address or domain name followed by the relative path to the requested resource from the web server's root folder (that is, the folder which holds the portion of the file system the web server has access to). If no port is specified, the client automatically sends HTTP requests to port 80, and HTTPS requests to port 443. The request additionally holds some headers, the type of the request, and optionally, some other content. There are these two important types from our perspective:
  - GET: In a GET request, everything is encoded into the URL. If a script is specified as the destination, the parameters are encoded in the URL as key-value pairs separated with a =. The start of the parameters are marked with ?, while the parameter separator is &. A GET request with a CGI script can look like the following: `http://mysite.com/script.php?param1=value1&param2=value2`. There is no per standard character limit on GET requests, but as they are basically URLs, using it for sending a very long representation of a complex geometry for example is impractical.

- POST: POST requests are exclusively used with CGI scripts. In the URL, only the destination is specified; the parameters are contained in the body of the request. POST requests leave no trace, therefore, they are good for sending sensitive data to the server (for example, authentication). They are also commonly used to send insensitive form data in bulk, or to upload files to the server.

POST requests can actually target static resources due to bad design. In those cases, however, it is up to the web server how it handles the request. Apache, for example, treats them as GET requests, returning the content of the static resource.

- **Response**: If a web server is listening on the specified server's specified port, it receives the request data. If a static resource is requested by a GET request, it simply serves it as is. If a CGI script is the destination resource, it parses the parameters specified in the URL or in the POST request's body, and supplies them to the CGI script. It waits for the response of the script, then sends that response back to the web client.

A single-server machine can host an arbitrary number of server applications listening on different ports. They can receive requests from appropriate clients simultaneously, and respond to them (*Appendix 1.16*).

Spatial servers

The only question that remains is; how can spatial data be inserted into this architecture? Well, they can be stored as static resources in vector formats like KML or GeoJSON. The browser can read the content of these structured text files, and client-side web mapping software can use them. Publishing raster layers is a little more tricky. As web browsers do not have a concept about raster data, they need those layers as regular images. Therefore, we have to create a representation model on the server side, and send the resulting images to the client. The usual way of storing pre-rendered raster layers is to tile them, and serve the tiles. Then, if the client-side web mapping software knows the tiling scheme, it can create an interactive map by requesting visible tiles, and sewing them together. For this, we need to create tiles for various zoom levels (fixed scales) for the entire extent of our raster layer. In order to improve compatibility, there are various open source tiling standards. Two of the more popular standards are OpenStreetMap's slippy map, and OSGeo's **TMS (tile map service)**.

The other way to serve maps is, of course, via a CGI application. QGIS Server, for example, is a completely valid CGI application we can use with any web server. By sending parameters in a URL, we can get an image containing the requested layers rendered by QGIS Server on the fly. To make this concept a little bit more complex, there is also GeoServer, which is written in Java. Java is an exceptional language for writing web applications, as web servers cannot invoke those software directly as CGI scripts. In order to use a Java web application, we need a Java servlet, which is a specialized web server for running it. The platform-independent binary version of GeoServer is bundled with Jetty, a lightweight Java servlet listening on port 8080:

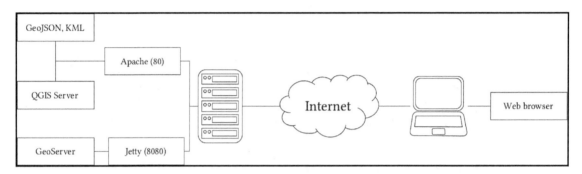

Using spatial servers instead of static files has a lot of advantages. For example, spatial servers can read out layers from spatial databases, therefore, we can always provide up-to-date data. We can also send only extracts of large datasets querying them by various means, or reproject them on the fly. As spatial servers need to be invoked by using some parameters, their interfaces are also standardized. Most of those standards are also maintained by experts and organizations forming the **Open Geospatial Consortium** (**OGC**). They define interfaces, that is, the communication between web clients and spatial servers. From the various OGC standards, there are some targeting the web. They are called **Open Web Services** (**OWS**), and define general spatial data transmission over the web. These standards are as follows:

- **Web Map Service (WMS)**: Layers rendered on the server side, and returned to the client as regular images.
- **Web Map Tile Service (WMTS)**: Layers rendered, tiled, and cached on the server side. Tiles are returned to the client.
- **Web Feature Service (WFS)**: Vector layers sent to the client as structured text. By default, WFS uses the GML vector format.
- **Web Coverage Service (WCS)**: Raster layers sent to the client as raw raster data. It is used by desktop GIS clients capable of reading raw raster data (for example, QGIS), therefore, we won't discuss it further.

While OWS standards allow communicating through GET and POST requests, both QGIS Server and GeoServer are mainly used with GET requests, therefore, parameterized URLs. There are some common parameters for every service, which are listed as follows:

- **Service**: The abbreviation of the requested service. It can be **WMS**, **WMTS**, **WFS**, or **WCS**.
- **Version**: The version of the requested service, as spatial servers can provide data using different versions for backward compatibility. For WMS, it is usually **1.3.0**; **1.1.0**, **2.0**, and **2.0.2** are widely used for WFS, while for WMTS, it is **1.0**.
- **Request**: The type of the operation that the spatial server should perform. A common value is GetCapabilities, which requests the metadata of the provided layers used with the requested service. For WMS, it is usually GetMap, for WFS it is GetFeature, while for WMTS, it is usually GetTile.

If we put this together, we can craft an URL, which can query a spatial server's WMS capabilities the following way:

```
http://mysite.com/spatialserver?
  Service=WMS&Version=1.3.0&Request=GetCapabilities
```

If you choose a service, there are other service-related parameters you have to provide to get spatial data or maps as output. As both of the servers we will use have convenient methods for creating previews, we won't discuss service-related parameters further. You can read more about them by downloading the whitepapers of the standards from OGC's website at http://www.opengeospatial.org/standards.

Using QGIS for publishing

QGIS offers a very easy and convenient way to publish QGIS projects with its own spatial server. It is the CGI application QGIS Server, which we have already configured in Chapter 1, *Setting Up Your Environment*.

If you are using Windows, and could not configure QGIS Server properly, don't worry, just skip to the GeoServer part (*Using GeoServer*).

Similar to the popular UMN MapServer, QGIS Server is a simple CGI application which does not track the published data. While MapServer needs a configuration file where paths to the data sources are defined along with other configuration parameters, QGIS Server needs a QGIS project file, which contains the paths along with other information, like styling. We can provide the project file's absolute path in a `map` parameter.

> While web servers can only access a portion of the file system, CGI scripts can access anything they have permission to read or write. Always consider this when using CGI scripts.

Let's craft a URL which queries the WMS capabilities of QGIS Server using one of our QGIS projects. As we are on the same machine as the server, we can use the placeholder `localhost` instead of a domain name or an IP address:

```
http://localhost/cgi-bin/qgis_mapserv.fcgi?
 map=/home/debian/practical_gis/
  ch10.qgs&Service=WMS&Version=1.3.0&Request=GetCapabilities
```

The response should look similar to the following:

```
      </LegendURL>
     </Style>
    </Layer>
   <Layer queryable="1">
    <Name>suitability</Name>
    <Title>suitability</Title>
    <CRS>CRS:84</CRS>
    <CRS>EPSG:23700</CRS>
    <CRS>EPSG:4326</CRS>
    <CRS>EPSG:3857</CRS>
    <EX_GeographicBoundingBox>
      <westBoundLongitude>17.5961</westBoundLongitude>
      <eastBoundLongitude>18.8541</eastBoundLongitude>
      <southBoundLatitude>45.7388</southBoundLatitude>
      <northBoundLatitude>46.4281</northBoundLatitude>
    </EX_GeographicBoundingBox>
    <BoundingBox CRS="EPSG:3857" maxx="2.09883e+06" minx="1.95878e+06" maxy="5.84923e+06" miny="5.73858e+06"/>
    <BoundingBox CRS="EPSG:4326" maxx="46.4281" minx="45.7388" maxy="18.8541" miny="17.5961"/>
    <BoundingBox CRS="EPSG:23700" maxx="634947" minx="538426" maxy="120428" miny="44800"/>
    <Style>
      <Name>default</Name>
      <Title>default</Title>
      <LegendURL>
       <Format>image/png</Format>
       <OnlineResource xmlns:xlink="http://www.w3.org/1999/xlink" xlink:type="simple" xlink:href="http://localhost/cgi-bin/qgis_mapserv.fcgi?
       map=/home/debian/practical_gis/ch10.qgs&&SERVICE=WMS&VERSION=1.3.0&REQUEST=GetLegendGraphic&LAYER=suitability&FORMAT=image/png&STYLE=default&SLD_VERSION=1.1
      </LegendURL>
     </Style>
    </Layer>
   <Layer queryable="1">
    <Name>mean_coords_dist</Name>
    <Title>mean_coords_dist</Title>
    <CRS>CRS:84</CRS>
    <CRS>EPSG:23700</CRS>
    <CRS>EPSG:4326</CRS>
    <CRS>EPSG:3857</CRS>
    <EX_GeographicBoundingBox>
      <westBoundLongitude>17.6031</westBoundLongitude>
      <eastBoundLongitude>18.8527</eastBoundLongitude>
      <southBoundLatitude>45.7404</southBoundLatitude>
      <northBoundLatitude>46.4214</northBoundLatitude>
    </EX_GeographicBoundingBox>
    <BoundingBox CRS="EPSG:3857" maxx="2.09867e+06" minx="1.95956e+06" maxy="5.84814e+06" miny="5.73885e+06"/>
    <BoundingBox CRS="EPSG:4326" maxx="46.4214" minx="45.7404" maxy="18.8527" miny="17.6031"/>
```

If you are using Windows, you might need to use the path `http://localhost/qgis/qgis_mapserv.fcgi.exe` to reach QGIS Server.

A long XML response shows if we have a working QGIS Server, which can access our projects. By installing QGIS Server, it automatically integrates itself in QGIS, and makes our projects publishable with WMS. To see what those published maps look like, we can use QGIS as a client. In order to load a WMS layer, we need to connect to a spatial server publishing WMS layers first, using the following steps:

1. Click on the **Add WMS/WMTS Layer** button on the left toolbar.
2. Click on **New** to define a new connection.
3. Name the connection (it can be anything), and provide the URL to the QGIS Server application with the map parameter pointing at a QGIS project file (for example, `http://localhost/cgi-bin/qgis_mapserv.fcgi?map=/home/debian/practical_gis/ch04.qgs`). I will use the carefully styled map from `Chapter 4`, *Creating Digital Maps*.
4. Click on **OK** to save the connection, and close the dialog.
5. Click on connect to see every layer published as a WMS map.

6. Select the topmost layer named after the project file to load every layer from the project. Add the whole map with the **Add** button:

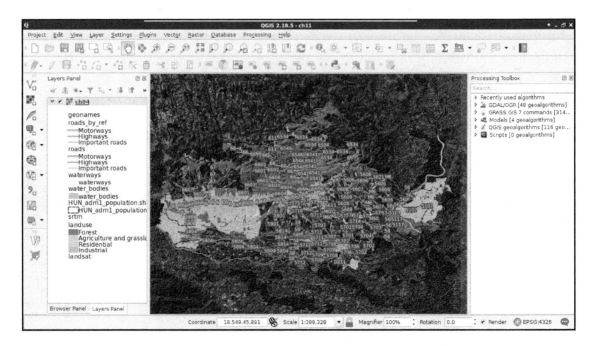

If you experience character encoding problems with labels, you have to set character encoding explicitly. To do that, open the project you are using with QGIS Server, open **Properties | General** on the source layer of the problematic labels, and set the **Data source encoding** from **System** to the correct value.

We can see the styled layers, although there are two very conspicuous problems--there are some layers published which we had disabled in our project, and it uses the projection EPSG:4326, making the map look distorted. The overall composition is not very aesthetic, which we can resolve by opening the project in QGIS:

1. Open **Project | Project Properties** from the menu bar.
2. Navigate to the **OWS server** tab.

As QGIS Server serves projects, we can configure the behavior of the QGIS Server in the project properties.

There are several general sections in the dialog which are useful for basic configuration, like the following:

- **Server capabilities**: Basic metadata about the provider. If enabled and filled out, that data is supplied with the capabilities of XML by the server application.
- **CRS restrictions**: QGIS Server can create WMS images in up to three CRSs by default. Two of them are the popular EPSG:4326 and EPSG:3857 CRSs. If the published project uses a third CRS, it can be also used by default. To customize the CRSs, WMS layers can be queried, we can enable **CRS restrictions**, and manage the allowed projections manually.
- **Exclude layers**: By default, every layer in the project is published as a WMS image. In order to exclude some of the layers, we can add them in the **Exclude layers** section.
- **WFS capabilities**: Spatial data is not published by QGIS Server in raw vector format by default. We can enable WFS layer-wise, by checking the **Published** box on the appropriate vector layers. We can also enable **WFS-T** (**WFS-Transaction**) on some layers (that is, **Update**, **Insert**, and **Delete**). The only requirement is that the layer has to be in a format QGIS can write in place.
- **WCS capabilities**: Similar to WFS, we have to enable publishing raw raster data with WCS layer-wise.

Do not hesitate to try out WFS and WCS on some layers. Once enabled, you can request WFS and WCS layers with the **Add WFS Layer** and **Add WCS Layer** buttons similar to adding new WMS layers.

For now, let's only exclude the disabled layers from the WMS service.

1. Add the disabled layers in the **Excluded layers** list with the green plus icon.
2. Apply the changes, and save the project.
3. Open a new project, or an existing one used for this chapter.
4. Add every layer from the previously saved project as a single WMS layer as we did previously. Before adding the layer, click on the **Change** button next to the default CRS's name (**WGS 84**), and select the local projection we use from the list.

If QGIS Server still provides the excluded layers, remove the layer, and add a new WMS layer. This time though, do not select the parent layer named after the project file, but select every child manually. You can do that by holding down the *shift* key while clicking on each layer you would like to include.

Now we have a stack of styled layers ready to use in a web application:

The image we see is the same as we will see on our webpage if we use the layers with web-mapping software. This is the greatest benefit of using QGIS Server--as it uses the same libraries as QGIS, the served images will be the same as we can see in the project's styled layers in QGIS.

Using GeoServer

Publishing data with QGIS Server is indeed very easy, although being a simple CGI application, it has limited capabilities. Unlike QGIS Server or UMN MapServer, GeoServer is a full-blown web application. That is, it has an internal data structure used for storing a lot of things including styles, authentication data (profiles), and references to raw spatial data. GeoServer is a Java web application, therefore, it needs a Java servlet to work (included in the default binaries). Therefore, if we start GeoServer with the technique described in Chapter 1, *Setting Up Your Environment*, we have to wait until the Java virtual machine starts up and the application initializes itself. Once GeoServer is running, we can access it from a browser by connecting to the 8080 port of our local server, and access the geoserver application as follows:

```
http://localhost:8080/geoserver
```

As GeoServer uses authentication, first we have to log in with the default admin credentials. The default admin username is `admin`, while the default password for this user is `geoserver`:

General configuration

As GeoServer is a full application with users and authentication, the first and most important thing to change is the master and admin passwords. The master password belongs to the superuser **root**, which is a fixed administrator user. Unlike the user admin, it cannot be removed, therefore, GeoServer remains manageable no matter if the admin account is accidentally removed:

1. Go to **Security** | **Passwords** from the navigation panel on the left-hand side.
2. Click on **Change password** next to the password provider.
3. Provide `geoserver` as the original password, and supply a new password for the **root** account.
4. Click on the **Change Password** button to apply the changes.
5. Go to **Security** | **Users, Groups, Roles**.
6. Select the **Users/Groups** tab, and click on the `admin` user.
7. Change the password by typing in a new password and confirming it in the corresponding fields.

8. Click on the **Save** button at the bottom of the page to apply the changes:

As we can see, there are three categories we can use in user management. There are regular users with user names, passwords, and individual permissions. There are groups, which can ease user management. A single user can be assigned to multiple groups. There is also a special category--roles. Roles are similar to groups, as a single user can use multiple roles. Using roles is significant in several aspects of permission management (for example, restricting access to services). In GeoServer roles are like responsibilities (e.g. admin, editor, user). Groups are grouping users and roles together so that common security combinations can be easily applied.

By default, GeoServer provides spatial data to any client-side request without authentication. As there is only one method in OWS capable of modifying the data sources (WFS-T), which is disabled by default, this can be considered safe. You can learn more about locking down services from the official guide at `http://docs.geoserver.org/latest/en/user/secu rity/service.html`.

The second thing we should configure is the metadata of the services. As GeoServer uses an internal data structure to store and provide spatial data added to it, metadata is configured for the entire application, not for individual projects. There are two kinds of metadata in GeoServer--global and service related. We can access global metadata that contains information describing the server's owner or maintainer by opening **About & Status | Contact Information**. Service-related metadata can be set by accessing the service settings under the **Services** section. Besides setting metadata, we can modify the behavior of the given service.

It is a good practice to disable services you do not intend to use, especially if you expose your GeoServer instance on the web. Raw data transmissions (that is, WFS and WCS) can generate a lot of traffic, and anybody can use them if they are available. You can disable services in the service configuration. For example, to disable WCS, you can open **Services | WCS**, and uncheck the **Enable WCS** box.

GeoServer architecture

When it comes to data sources, GeoServer offers a hierarchical structure that consists of workspaces, stores, and layers. GeoServer's architecture is simple enough to understand quickly, still, there are some tricky parts that need to be discussed:

- **Workspace**: A group that contains different elements, like stores, layers, styles, or workspace-specific configurations. This is similar to a QGIS project file.
- **Store**: A connection between spatial data stored on the disk and GeoServer. A single store can contain multiple layers if the storage type is capable to do so (for example, PostGIS).
- **Layer**: A published layer from an existing connection (store). Unlike QGIS Server, GeoServer does not publish every layer from the defined source; we have to explicitly select and publish our layers of interest:

If we go through **Data** | **Workspaces**, **Data** | **Stores**, and **Data** | **Layers**, we can see that every default store is put into a workspace, and layers from that store inherit the name of the workspace. Putting stores into a workspace is mandatory. Using a workspace is only optional for styles. This is because if we put a style into a workspace, it can only be used for layers in the same workspace, otherwise, it can be used globally. Workspaces behave similar to schemas in PostgreSQL. Layers can have the same name in different workspaces, and if we access layers globally, we should refer to them with their fully qualified names in the form of `workspace:layer`.

GeoServer offers several ways to access its services. The most general way is to use its global endpoint `ows` in the following format:

```
http://localhost:8080/geoserver/ows
```

By using this endpoint, we can supply every OWS parameter required for a valid request. For example, to get the WMS capabilities of GeoServer, we can use the following URL:

```
http://localhost:8080/geoserver/ows?
  Service=WMS&Version=1.3.0&Request=GetCapabilities
```

GeoServer has quite advanced rewrite rules, therefore, we can specify the service name after, or instead of `ows`. Thus, to get the same result, we can write the request the following ways:

```
http://localhost:8080/geoserver/ows/wms?
  Version=1.3.0&Request=GetCapabilities

http://localhost:8080/geoserver/wms?
  Version=1.3.0&Request=GetCapabilities
```

Alternatively, if we would like to only access the content of a single workspace, we can use a virtual endpoint by including the name of the workspace in the URL as follows:

```
http://localhost:8080/geoserver/topp/ows?
  Service=WMS&Version=1.3.0&Request=GetCapabilities
```

The tricky part comes while defining custom service behavior to workspaces. As we could see from the capabilities of GeoServer's WMS service, it can provide WMS images in every known CRS. In order to demonstrate this custom behavior, let's restrict the CRSs of the **topp** workspace to `EPSG:4326` and `EPSG:3857`:

1. Go to **Data | Workspaces**, and select the **topp** workspace.
2. Check the **WMS** box under **Services** to enable workspace-specific WMS configuration for this workspace. Click on the **Save** button to apply the changes.
3. Go to **Services | WMS**, and select the **topp** workspace in the **Workspace** menu.
4. Find the **Limited SRS list** field, and supply the EPSG codes of the two projections (`4326, 3857`). Save the edits by clicking on **Submit** at the bottom of the page.

Let's connect to the global endpoint of GeoServer, and query its WMS capabilities (`http://localhost:8080/geoserver/ows?Service=WMS&Version=1.3.0&Request=GetCapabilities`). Next, query its capabilities using the **topp** virtual endpoint (`http://localhost:8080/geoserver/topp/ows?Service=WMS&Version=1.3.0&Request=GetCapabilities`). As you can see, the virtual endpoint has now limited CRS capabilities, while the global endpoint can still provide layers from the **topp** workspace in any projection GeoServer knows. This is the tricky part of managing workspaces. As the global endpoint has to provide every published layer with every service enabled in the global configuration, it cannot use workspace-specific configuration. On the other hand, when using a virtual endpoint, GeoServer will try to use configuration for the corresponding workspace. If it cannot find any, it simply falls back to the global service configuration.

If you would like to have workspace-specific configuration only, you have to enable the desired services in every workspace, configure them, then disable the services in the global configuration.

Adding spatial data

Before adding some data to GeoServer, let's create a new workspace for it. This way, the data will be safely separated, and we will be able to access the content by using a virtual endpoint:

1. Open **Data** | **Workspaces**.
2. Select **Add new workspace**.
3. Specify a name for the workspace in the **Name** field.
4. Give a namespace for the workspace in the **Namespace URI** field. The namespace must be a URI (that is, a URL or a URN). It does not have to point to an existing resource, only its uniqueness is what matters (compared to the URIs of the other workspaces). For example, a URL can be `http://practical-gis.com/geoserver`, while a URN can be `urn:practical-gis:geoserver:`

There is a great article about URIs at `https://blog.4psa.com/url-urn-uri-iri-why-so-many/`.

Now we can start defining stores for our data sources. GeoServer does not create internal copies of the input data. It only uses references to the input files or databases, and reads the underlying spatial data on demand:

1. Go to **Data** | **Stores** and select **Add new Store**.

   Although GeoServer has somewhat limited knowledge about spatial formats, it still offers stores for the three formats we used previously in this book-- `Shapefile`, `PostGIS`, and `GeoTIFF`.

 One of the advantages of GeoServer is its modular architecture. As it is written in Java, which is an interpreted language, you can extend GeoServer's capabilities with custom Java code. There are some official extensions downloadable from `http://geoserver.org/release/stable/`. In the **Extensions** section, you can download various vector and coverage format extensions. You can install an extension by extracting the Java archives to GeoServer's `webapps/geoserver/WEB-INF/lib` folder.

2. Select **PostGIS**.
3. Set the **Workspace** to the one we created previously.
4. Give a name to the data source. The name won't be used by the services, although it must be a unique one.
5. Provide the name of the database in the **database** field. If you followed the book's naming conventions, it is **spatial**.
6. Provide the name of the schema containing spatial tables in the **schema** field.

7. Provide the username, and optionally, the password. Although it should be safe to use a PostgreSQL role with write privilege, it is a good practice to use double protection if we do not intend to write PostGIS tables from GeoServer. Therefore, we should use the public role we created with the corresponding password. If you are on a Unix system, the **password** field can be left blank:

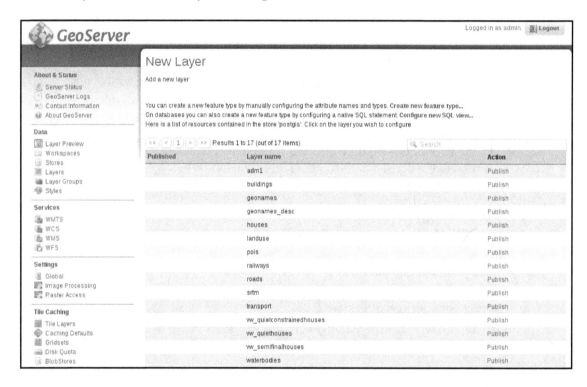

As you can see, if we use a store which can provide multiple layers, we can access every layer by their names. On the other hand, if the store is a single file which can contain only one layer, we can only access that single layer from the specific store. Before publishing a layer, let's add three more single file stores.

8. Open **Add new Store** from **Data** | **Stores** again.

9. Select the **Shapefile** option.

10. Select the workspace we created for our data.

11. Give a name to the store.

12. Browse out the shapefile containing suitable areas from our MCE analysis in the **Shapefile location** field by clicking on **Browse**.

13. Select the appropriate character encoding. Probably, it is UTF-8.

14. Save the store with the **Save** button, and return to the **Add new Store** page.

15. Select the GeoTIFF format, and choose the correct workspace.

16. Give this raster store a name, and browse out the suitability layer from the results of the MCE analysis.

17. Save the store with the **Save** button.

18. Add the clipped SRTM layer in the local projection we used with QGIS the same way as the suitability layer.

As you can observe, GeoServer offers a data directory it can instantly access in its file browser. It is the data_dir folder located in GeoServer's directory. If you create a folder there, and copy the data sources there, you can ease browsing. Furthermore, if you are planning to run a production server, you most likely would like to remove example workspaces and datasets. After removing them from GeoServer, you can free up some space by removing them physically from the data and coverages folders inside the data_dir folder.

As we now have our data sources configured, we can start publishing relevant layers. Let's start with some of our vector layers:

1. Open **Data | Layers**.
2. Select **Add a new layer**.
3. Select one of our newly defined stores (for example, suitable areas).
4. Find a layer to publish, and click on its **Publish** option.
5. Supply a name for the layer if the default one is not appropriate.
6. Look at the **Coordinate Reference Systems** section. GeoServer will try its best to find out the CRS of the layer in the **Native SRS** field. If it can successfully find the corresponding SRID, it automatically fills out the **Declared SRS** field (that is, the default CRS of the published layer). If it cannot, we should know the EPSG code of the layer's CRS. In this case, we have to provide it in the **Declared SRS** field, and leave the **Force declared** option selected. This way, GeoServer will not care about the original CRS of the layer; it will apply the one we provided on it.
7. Calculate the bounds of the layer automatically by clicking on the **Compute from data** option in the **Bounding Boxes** section.
8. Calculate the `WGS84` bounds of the layer by clicking on the **Compute from native bounds** option.
9. Publish the layer by clicking on the **Save** button.
10. Repeat the steps for the other required vector layers. We will need the administrative boundaries, the GeoNames layer, the land use, the roads, the waterways, and the water bodies.

Now we have a published layer, which we can preview easily from GeoServer. The only thing we have to do is to go to **Data | Layer Preview**, and select the **OpenLayers** option in the row of the newly published layer. The result shows what the layer will look like when accessed through GeoServer's WMS service. Although GeoServer uses OpenLayers for creating its previews, the map's content will look the same in any other web-mapping application.

 OpenLayers is a web mapping library for creating interactive maps on the client side. We will talk about web mapping in `Chapter 13`, *Creating a Web Map*.

The suitable areas layer in the preview window should look like the following:

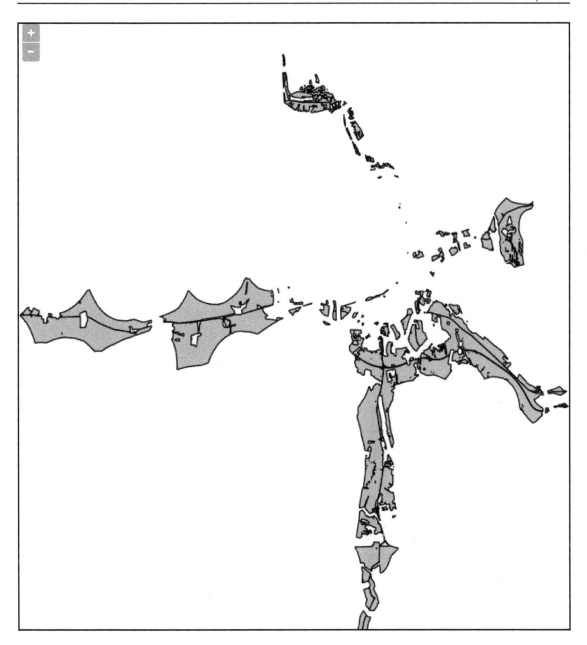

You can also use QGIS to preview layers by connecting to one of the endpoints of GeoServer.

It's time to publish some raster layers. Publishing raster layers is a little different, but only requires one extra consideration:

1. Open **Data | Layers**, add a new layer, and select the suitability layer from its store.
2. The bounding box is automatically calculated this time, however, the default option for reprojecting is **Reproject native to declared**. It is completely useless when the native and declared SRSs are the same, while it is harmful if GeoServer cannot identify the CRS of the raster data correctly. Set it to **Force declared**.
3. Repeat the steps for the SRTM raster.

 Don't worry about the preview of the raster layers. We will fix that in the next chapter by applying custom styles to them.

Now we can require individual layers, but how can we create compositions? Well, of course, we can define multiple layer names in a single `GetMap` request, although this approach can be quite inconvenient, especially, when we would like to stack a lot of layers. On the other hand, GeoServer has the capability of grouping layers to form a server-side composition. The only limitation of this approach is that it can only be used for WMS requests with layers in a single workspace:

1. Open **Data | Layer Groups**.
2. Select the **Add new layer group** option.
3. Name the layer group, give a descriptive title, and select the workspace we are working with.
4. Supply the `EPSG` code of our local projection in the **Coordinate Reference System** field (for example, `EPSG:23700`).
5. One by one, add every layer from the road map composition we made in QGIS earlier using **Add Layer**.
6. Click on **Generate Bounds** to compute the bounding box of the composition automatically.
7. Define the correct layer order by using the green arrows in the layer list. The list defines the drawing order, therefore, the first item is drawn at the bottom, while the last layer will be at the top.
8. Save the composition with the **Save** button.

If we preview the newly created layer group, we can see the raw composition, that is, every component with their default styling stacked on each other. Now it is only a matter of styling to get a similar map like in our QGIS project:

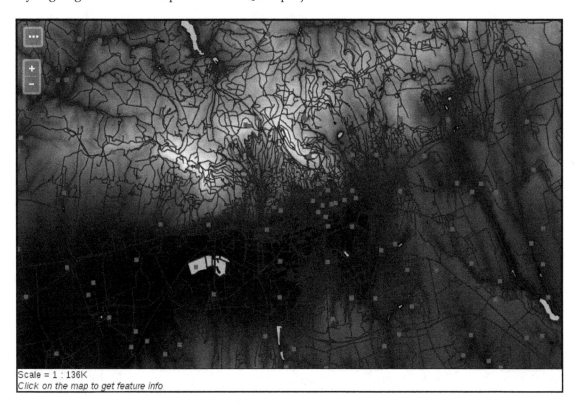

Tiling your maps

One of the greatest perks of GeoServer is its internal tiling and tile-caching capability. It uses GeoWebCache, which is integrated into GeoServer, and enabled by default. Of course, this behavior can be also considered a downside, as cached tiles always use up some disk space. GeoServer's default tiling behavior is dynamic. It creates tiles on demand, and stores them until expiration for reuse. Tiling can greatly increase the speed of serving images, with the expense of additional disk usage. Although there are multiple tiling services served by GeoServer, every one of them can be served with the same tiles (gridset). Only the layout differs, which is calculated by GeoServer before serving the right tiles.

If we go to **Data | Layers**, and inspect a layer by clicking on its name, we can see the default tiling and tile caching options in the **Tile Caching** tab. As we can see, tile caching is enabled for two formats (PNG and JPEG) by default. Furthermore, tiles can be generated for two gridsets, one using the EPSG:4326 CRS, and one using the EPSG:9009013 CRS. Finally, tiles can be generated for every style associated with the given layer, which is only the default style by default.

EPSG:900913 is the historical code for the Web Mercator (EPSG:3857). As Google Maps created and used it first, and EPSG refused to accept it as a valid CRS for a long time, Christopher Schmidt coined the code name of 900913 (GOOGLE if the numbers are rotated by 180 degrees individually), and it got popular wrongly as EPSG:900913.

If we go into **Tile Caching | Gridsets**, we can open the properties of the two enabled gridsets, and check their properties. Gridsets tile up the whole extent of a CRS, and create a layout for every zoom level. They start with only a few tiles for the smallest zoom level, and increase quadratically with every defined new zoom level. If we calculate the theoretic maximum of stored tiles for the EPSG:4326 gridset, we get the value 11,728,124,029,610. For the other gridset, we get a much higher value of 1,537,228,672,809,129,200. These are the number of tiles which can be stored by GeoServer, theoretically, for both of the image formats. According to GeoSolution's presentation at https://www.slideshare.net/geosolutions/geoserver-in-production-we-do-it-here-is-how-foss4g-2016 (they are contributors to the GeoServer project), GeoWebCache's tile storing mechanism is very efficient, as about 58,377 tiles can fit into a single megabyte. Still, to store a layer in both of the default gridsets in a single format, we would need about 24 exabytes of space. I do not think we would need any more proof that managing tile caching is a very good practice, otherwise, GeoWebCache can fill up every bit of free disk space it can use quite quickly.

You can alter the default caching behavior for new layers by navigating to **Tile Caching | Caching Defaults**, and altering the corresponding options. By unchecking the **Automatically configure a GeoWebCache layer for each new layer or layer group** option, GeoServer won't make a cached layer for new layers automatically. Similarly, you can alter the default formats for default data types, and the default gridsets used by GeoWebCache.

The first thing we can configure is the disk quota. Despite the large space requirement of tiles, we shouldn't give up on serving tiled variants of WMS images, as they are beneficial if used wisely. We can restrict the space GeoWebCache can take up, and let it create tiles for whatever layers we would like to give a boost. If it reaches its quota, it will free up some space by removing or overwriting unused tiles:

1. Go to **Tile Caching | Disk Quota**.
2. Check in the **Enable disk quota box**.
3. Specify an appropriate size for GeoWebCache in the **Maximum tile cache size** field.
4. Select a recycling behavior from the two available options (that is, **Least frequently used** and **Least recently used**).
5. Apply the disk quota by clicking on **Submit**.

Now we can deal with the tiled variants of our layers. Despite having a quota, we should not leave GeoWebCache filled with tiles we do not need. We can manage tile layers by going to **Data | Layers**, opening a layer, and navigating to the **Tile Caching** tab like we did previously:

- If we would like to disable tile generating for the entire layer, we can uncheck the **Create a cached layer for this layer** option.
- If we would like to disable only caching tiles for a layer, we can uncheck the **Enable tile caching for this layer** option.
- We can disable the `image/jpeg` format for every layer safely. The usual image format we use on the web is PNG, as it can store transparency. By using JPEG, we get smaller image sizes, but we also get a white background where there aren't any features.
- We can also remove unused gridsets from layers with the red minus button. For the gridsets that we would like to use, we can define the minimum and maximum zoom levels we would like to publish or cache.

You can go to **Tile Caching | Tile Layers**, and remove layers from there. You achieve the same effect as turning off the **Create a cached layer for this layer** option, and you can also bulk-remove tiled versions of layers.

As our GeoServer is not public, do not bother with the tile setup for now. Let's create a tiled variant for our new layer group in our local projection instead:

1. Go to **Tile Caching | Gridsets**.
2. Select the **Create a new gridset** option.
3. Name the gridset to represent the local projection we use. Avoid using special characters, whitespaces, or slashes.
4. Supply the EPSG code of the local projection in the **Coordinate Reference System** field.
5. Click on **Compute from maximum extent of CRS** to make GeoServer calculate the bounding box of the gridset automatically.
6. Click on **Add zoom level** as many times as the zoom levels we would like to provide. Create at least 10 zoom levels. The optimal number of zoom levels highly depends on the size of the CRS's extent. You can take a hint from the Scale column of the zoom levels. A scale of 1:500 is building level.
7. Save the new gridset with the **Save** button.
8. Navigate to **Data | Layer Groups**, and select our layer group from the list. This is the same as selecting the layer group from the layers list, but more convenient.
9. Go to the **Tile Caching** tab, and add our new gridset by selecting it in the **Add grid subset** field, and clicking on the green plus button.
10. Save the edits made to the layer group.

Now we can preview the tiled version of our layer group in our CRS by navigating to **Tile Caching | Tile Layers**, finding the row of our group layer, and selecting the appropriate gridset and format combination from its Preview column:

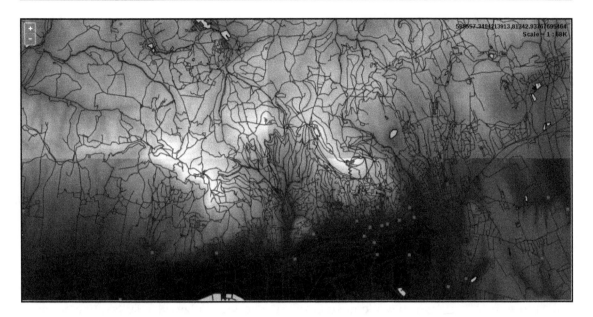

 Do not worry about those sharp tone changes in the DEM layer. GeoServer, by default, stretches the local minimum and maximum values of a subset of the DEM to the grayscale color space at a time, not the global min/max values. We will fix that by applying a palette with predefined intervals in the next chapter.

We can instantly see the greatest benefit of using tiled layers when we browse the preview. The first time when we pan or zoom around, GeoServer takes some time to render the tiles. Then, if we navigate to already visited areas, it loads the content instantly. If we navigate to **Tile Caching | Disk Quota**, we can also see our disk slowly filling up by browsing the preview. You may ask now: where are the tiles? They are stored in GeoServer's `data_dir/gwc` folder. Tiling provided by GeoWebCache does not only mean that a separate module does tile providing and caching; it also means that we can only request tiled resources from a third endpoint--`gwc`:

```
http://localhost:8080/geoserver/gwc
```

As GeoServer provides various tiling services from which only WMTS is an OGC standard (therefore, has similar parameters to WMS, WFS, and WCS providers), tile requests are slightly different. We have to specify the service in the path of the URL, and can use service-related parameters after that. For example, to query the WMTS capabilities of our GeoServer, we can use the following URL:

```
http://localhost:8080/geoserver/gwc/service/wmts?
 Version=1.0.0&Request=GetCapabilities
```

Summary

In this chapter, we gained some basic but essential understanding about the architecture of the web. We learned how to easily provide spatial data with QGIS Server. We also learned some of the basic principles of GeoServer, and how we can create a starting configuration, which we can later expand by gaining additional experience with the system. We discussed standardized OGC services called OWS, and how they work in practice. We managed to not only add some of our own spatial data to GeoServer and visualize the results, but also see how we can tile them, speeding up the server's response.

In the next chapter, we will learn about styling spatial data in GeoServer. Styling is not only one of the corner points of publishing spatial data, but also a weak point of GeoServer, as it offers way more possibilities than documentation. Creating styles can look cumbersome, fiddling, and way too scary to get into. We will see how we can better understand SLD styling, and what other, more efficient tools we can get our hands on.

12
Styling Your Data in GeoServer

In the previous chapter, we covered the basics of sharing content over the web. We discussed how the web works, and how we can send spatial data over it. Then, we set up QGIS Server to see how spatial data are rendered as regular images and visualized in a client. Then we went on and configured GeoServer to have a reliable spatial server in our service, even capable of tiling up rendered images, and not only providing, but also caching those tiles.

In this chapter, we will learn about styling vector and raster data in GeoServer. We will cover the basic symbolizers we can use, and the syntax of the style language used by GeoServer--**SLD (Styled Layer Descriptor)**. After you understand how SLD works, we'll go on and study the more convenient GeoServer CSS, which is a concise, **CSS (Cascading Style Sheet)**-based language available in GeoServer through an extension.

In this chapter, we will cover the following topics:

- Vector and raster symbology in GeoServer
- Writing simple SLD styles
- Creating styles easily with CSS

Managing styles

In GeoServer, we can manage style items the way we managed layers. Unlike layers, we can make styles global by not assigning them to a single workspace. This makes them usable with layers in different workspaces. On the other hand, as styles can be very specific (that is, they can use rules based on attribute data), we can also make them local by assigning them to a workspace. The styling scheme of GeoServer is very similar to QGIS and any other GIS software. We can use point, line, polygon, and raster symbolizers to describe visual properties.

These symbolizers can be explained as follows:

- **Point**: A symbol bound to a pair of coordinates. The symbol can be an image, or any other regular shape (for example, a square, triangle, or circle). If the symbol is a shape, it can have a stroke and a fill.
- **Line**: A linear symbol described with a stroke width and a stroke color.
- **Polygon**: A symbol applied to an area described by coordinate pairs in the CRS of the map. It has a stroke with a stroke width and a stroke color. Additionally, it has a fill, which can be described by a simple color, a color gradient, or a texture.
- **Text**: A rendered label that shows an attribute of the underlying feature. It can be described with common font properties usable with WYSIWYG word processors (like LibreOffice Writer).
- **Raster**: A rule or a set of rules which assigns colors to raster values.

Of course, styles only affect image outputs (like WMS or WMTS). There are numerous styles shipped with GeoServer. When we publish a layer, we have to assign a default style for it. If we skip this step, GeoServer automatically assigns an internal style for the given layer based on its type. Layers can have multiple styles. In a query, we can provide a style parameter, which requests a style name other than the default style of the requested layer. If the requested style exists, and it is assigned to the requested layer in GeoServer, it provides images styled according to the request.

We can see GeoServer's default styles by navigating to **Data** | **Styles**. As you can see, the default styles are global, therefore, we can use them with our layers if they fit. Let's open the style named **simple_streams** by clicking on it. The window we opened is the style editor, where we have access to these four style management tabs:

- **Data**: We can specify the attributes of the given style here. We can rename the style, or restrict it to a specific workspace. If we create a new style, we can also choose a format and generate some random template.
- **Publishing**: We can access all of our layers here and assign the style as default, or just associate it with them. Associated styles can be queried by using a style parameter as discussed before.
- **Layer Preview**: We can preview the style on any layer in this tab. By default, a layer using the given style is shown in the preview. We can change that by clicking on the previewed layer's name. This is a very convenient editing tool, as we can follow our changes visually.
- **Layer Attributes**: We can inspect the attribute data of the previewed layer here, which is useful if we use some attribute-based rules in our style.

The long and intimidating text in the style editor field is the style definition of the style we opened. As you can see, the style editor is accessible in every tab, making editing very convenient. There are also some of these following useful tools in the form of buttons, which we can use with our modified styles:

- **Validate**: To validates the style that we create anytime--GeoServer parses the draft, and points out any error it can find
- **Apply**: To save the modified style in place
- **Submit**: To save a newly created style first time, making it available to any layer
- **Cancel**: To go back and discard any change

Let's play around a little with this style to get a hang of the process:

1. Navigate to the **Layer Preview** tab
2. Click on the previewed layer's name (**sf:streams**), and select our waterways layer

Do not worry about the WGS 84 CRS of the previewed layer. In the preview map, GeoServer always renders the previewed layer using EPSG:4326 even if it is disabled in the list of CRSs usable by the WMS service.

3. In the style editor, change the stroke width to 1 pixel by modifying the `<ogc:Literal>2</ogc:Literal>` line. Optionally, change the color of the stroke to the color we used in QGIS (`#a5bfdd` for me) by modifying the `<ogc:Literal>#003EBA</ogc:Literal>` line.
4. Validate the change by clicking on the **Validate** button.
5. Save the modification by clicking on the **Apply** button. If the preview does not change immediately, pan or zoom the map a little bit.
6. Change to the **Publishing** tab.

7. Specify this style as the default style on the waterways layer by searching it, and checking the Default box in its row:

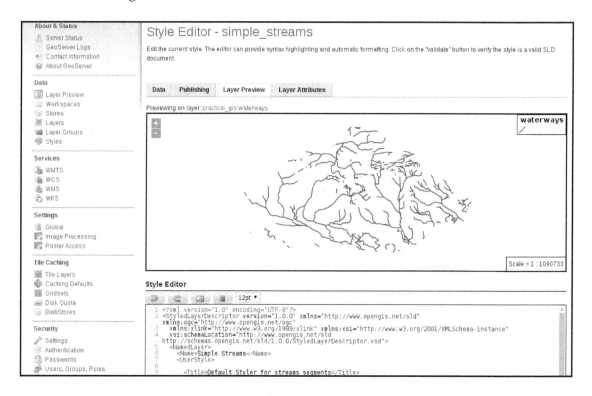

Writing SLD styles

GeoServer's renderer uses, and is basically built around, the SLD (Styled Layer Descriptor) specification. SLD is an **XML (Extensible Markup Language)** extension specified by OGC. It allows rich representation of the underlying spatial data regardless of their type. As SLD is XML based, it is a structured text using tags, attributes, and values. In order to understand SLD better, let's go through the description of the style we opened before.

The original SLD specification is very universal and extensive, therefore it is complex. Moreover, GeoServer has a great implementation capable of harnessing most of its features. Therefore, in order to keep the learning curve calm, we will only use a small subset of SLD. We have to make two assumptions in order to do this--we only use catalog styles (that is, static styles managed from GeoServer), and every vector layer contains only one type of geometry.

The first line in grey describes that the content as a valid XML document of version 1.0 with a UTF-8 character encoding. In the other lines, we can see the body of the SLD. As in every valid XML document, most of the elements have an opening (`<TagName>`) and a closing (`</TagName>`) tag. Between the two tags, the content of the element is described, whether it is a simple value, or a group of other elements. The body tag of an SLD document is `<StyledLayerDescriptor>`, which we can see in the second, and the last line. Opening tags can also have some attributes if applicable (for example, `<StyledLayerDescriptor version="1.0.0">`). The attributes of the `<StyledLayerDescriptor>` element are static, and out of concern, as they will be automatically generated if we generate a template for our new style. There are some additional elements we should understand to create a basic style. They are as follows:

- `<Name>`: The name of the parent element. It is mandatory for the `<StyledLayerDescriptor>` element, but not relevant (that is, it can be anything). It can be also used to describe some other elements optionally.
- `<UserStyle>`: A mandatory child of the `<StyledLayerDescriptor>` element, which contains the style definition.
- `<FeatureTypeStyle>`: At least one of this element should be present in a `<UserStyle>` element. It groups different rules of styling applied in a single rendering pass. Multiple elements of this kind can achieve complex styles.
- `<Rule>`: A single set of rules grouping filters and corresponding symbology.
- `<ogc:Filter>`: A filter element in a `<Rule>` element describing a subset of the vector layer which should be styled in the rule.
- `<PointSymbolizer>`, `<LineSymbolizer>`, `<PolygonSymbolizer>`, `<TextSymbolizer>`, `<RasterSymbolizer>`: Elements residing in a `<Rule>` element describing a single symbol of the underlying spatial data in the rule.
- `<Fill>`, `<Stroke>`, `<Label>`, `<Font>`, `<Mark>`, `<ColorMap>`: Children of the symbolizer elements describing the appropriate properties of the parent symbol. For example, a `<PolygonSymbolizer>` can have `<Fill>` and `<Stroke>` children, but cannot have a `<Mark>` child, as it describes a point symbol.
- `<CssParameter>`: An element describing a single parameter of a symbol's aspect. For example, a `<Stroke>` element can contain a `<CssParameter name="stroke">` element describing the stroke's color, and a `<CssParameter name="stroke-width">` element describing its width.
- `<ogc:Literal>`: An element containing a value. It is usually used as a child of an `<ogc:Filter>` element to provide a value.

Styling vector layers

Let's start by styling some vector layers that we have. First of all, as we have a river style, we should style our water bodies accordingly. Water bodies should have an outline with the same color as our rivers, and a fill with a lighter shade of blue. As we already have a river color, we can simply copy it out and apply to our new style. To find a lighter shade of blue, we can use an online color picker widget. There is a great widget created by Google, accessible by simply searching for `html color picker` in Google Search:

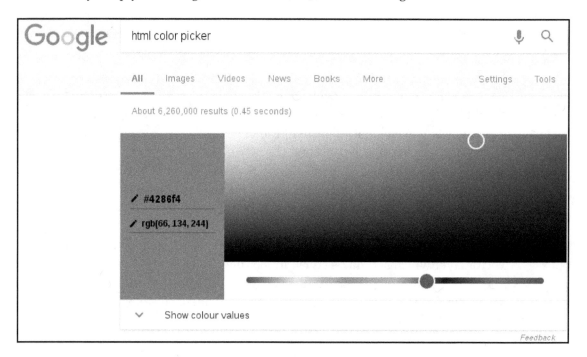

Styling waters

Let's style our waters by editing an existing style as follows:

1. Note down the color of the rivers in its SLD document (`#003EBA`).
2. Go to **Data | Styles**.
3. Initialize a new style by clicking on **Add a new style**.
4. Give it a short, but descriptive, name, like `waterbodies`.

5. Generate a polygon template by selecting **Polygon** in **Generate a default style**, and clicking on **Generate**.

6. Click on **Submit** to save the template.

7. Select the new style from the existing styles. Now we have access to the **Layer Preview** tab. Navigate to it, and select our water bodies layer as the preview layer.

8. Edit the stroke, and fill colors of the template. Optionally, edit the stroke width. Click on **Apply**, and check the result. Modify the fill color until you get a decent result.

9. Go to the **Publishing** tab, and set this style as a default for our water bodies layer:

Optionally, you can style the water bodies like we did it in QGIS. To do that, remove the `<Stroke>` element with its content, and set the fill color to the same shade of blue as the rivers.

Styling polygons

The last part was very simple. Next, let's style our administrative boundaries. From that layer, we only need our study area visualized with a dashed outline:

1. Create a new style, and name it appropriately.
2. Choose our workspace as **Workspace;** as this style will use an attribute from a layer, it won't be applicable to other layers.
3. Generate a default polygon style.
4. Save the template with the **Submit** button, open it again, and preview it on the administrative boundaries layer.
5. Remove the `<Fill>` element and its content from the SLD document, as we don't need a fill. Don't forget to remove the closing tag (`</Fill>`).
6. Add a third `<CssParameter>` element to the `<Stroke>` element of the style. It should define a dash, which can be achieved by using the `name="stroke-dasharray"` attribute. The dash array is a set of numbers separated by whitespaces. The first number describes the first dash's width in pixels, the second number is the first space's width, and so on:

   ```
   <CssParameter name="stroke-dasharray">5 5</CssParameter>
   ```

7. Filter the layer to only show our study area. This is a simple comparison, which needs an `<ogc:Filter>` element, an `<ogc:PropertyIsEqualTo>` element in it, a column name in an `<ogc:PropertyName>` element, and a value in an `<ogc:Literal>` element. We can put the code together by correct nesting, and get something similar to the following:

   ```
   <ogc:Filter>
    <ogc:PropertyIsEqualTo>
     <ogc:PropertyName>name_1</ogc:PropertyName>
     <ogc:Literal>Baranya</ogc:Literal>
    </ogc:PropertyIsEqualTo>
   </ogc:Filter>
   ```

8. Save the style, and declare it as the default style for the administrative boundaries layer.

We used some parameters that we have not not discussed before. You must be wondering how should you know the exact parameters to use in this case. The answer is very simple: you shouldn't. You can read out the parameters GeoServer can accept for different SLD elements from its SLD reference available at `http://docs.geoserver.org/stable/en/user/styling/sld/reference/index.html`.

Creating labels

Finally, let's create a style for our GeoNames layer showing only labels for the seats of the administrative divisions:

1. Create a new style, and name it appropriately. Restrict it to our workspace.
2. Generate a template for point geometries.
3. Save the template, open it again, and preview it on the GeoNames layer.
4. Optionally, edit the metadata to describe our use case.
5. Apply a filter in the `<Rule>` element filtering seats of administrative divisions. The filter we used in QGIS is `"featurecod" LIKE 'PPLA%'` or `"featurecode" LIKE 'PPLA%'` depending on the format we used when we saved the GeoNames extract. Remember: Shapefiles have a limit on the maximum length of a column name, therefore if we exported our GeoNames layer to a Shapefile previously, `featurecode` got truncated to `featurecod`. This is the expression we have to translate to an SLD filter. We can use a `LIKE` expression in SLD by using the `<PropertyIsLike>` element, although it is a little bit tricky. We have to provide three attributes to the element: `wildCard` for the wildcard character substituting any number of characters, `singleChar` for the wildcard substituting a single character, and `escape` for the character escaping a wildcard character. In the end, the filter should look similar to the following:

```
<ogc:Filter>
 <ogc:PropertyIsLike wildCard="%" singleChar="_" escape="\">
  <ogc:PropertyName>featurecod</ogc:PropertyName>
  <ogc:Literal>PPLA%</ogc:Literal>
 </ogc:PropertyIsLike>
</ogc:Filter>
```

6. Instead of using a point symbolizer, use a text symbolizer to display labels only. To achieve this, first remove the `<PointSymbolizer>` element with its content.

7. Add a `<TextSymbolizer>` element in the `<Rule>` element. The text should have a font type (for me, it is DejaVu Sans), a white color, and a black halo around it. For the font size, the default value of 10 pixels is a little bit small, therefore, we should use a somewhat bigger size. Furthermore, it should show labels from the `name` property of the layer, which we can describe in a `<Label>` element. In the end, we should get a symbolizer like the following:

```
<TextSymbolizer>
 <Label>
  <ogc:PropertyName>name</ogc:PropertyName>
 </Label>
 <Font>
  <CssParameter name="font-family">DejaVu Sans</CssParameter>
  <CssParameter name="font-size">12</CssParameter>
  <CssParameter name="font-style">normal</CssParameter>
 </Font>
 <Halo>
  <Fill>
   <CssParameter name="fill">#000000</CssParameter>
  </Fill>
 </Halo>
 <Fill>
  <CssParameter name="fill">#FFFFFF</CssParameter>
 </Fill>
</TextSymbolizer>
```

8. Validate the style. If it does not have any errors, save it, and declare it as the default style for the GeoNames layer.

Let's see how our layer group changed due to these SLD styles. First of all, we need to change its styling. As a layer group can have different styling than the components individually, we have to explicitly set the styles applied to the members of the group:

1. Go to **Data | Layer Groups**, and select our composition.
2. Check the **Default Style** box for the items we created a style for.
3. Save the changes by clicking on the **Save** button.
4. Go to **Data | Layer Preview**, and preview our layer group by clicking on **OpenLayers** next to it:

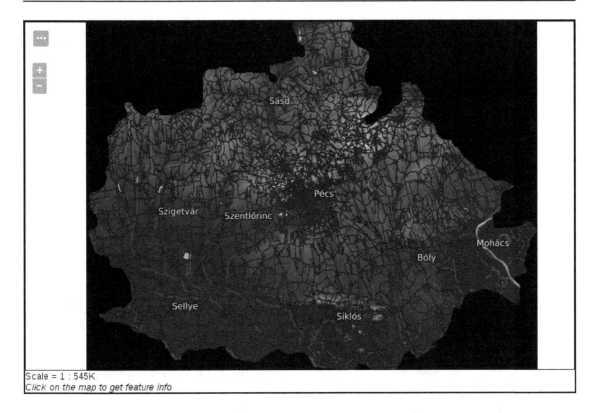

Scale = 1 : 545K
Click on the map to get feature info

Styling raster layers

When working with rasters, we have much less options in GeoServer, just like in QGIS. For singleband rasters, like our DEM, we can create a color ramp on which GeoServer interpolates the values of the raster data. We can define breakpoints on the color ramp with colors and raster values. In order to do this, first we need the minimum and maximum values of our SRTM DEM:

1. Open QGIS, load the SRTM DEM which is used in GeoServer, and note down its minimum and maximum values. We can use **Properties** | **Style** with the accurate min/max values, or we can read out the values from the **Metadata** tab.

2. Create a new style in GeoServer, and name it appropriately.

3. Generate a default raster style, and save the template. Open it again, and preview it on the DEM layer.

4. The preview shows nothing. This is a bug in the SLD generator, as it generates a `<FeatureTypeName>` element, which is not applicable on raster layers. Remove that element, and save the template.

5. In order to create a grayscale DEM, we need to interpolate the raster values between the black and the white colors. We can do so by creating a `<ColorMap>` element in the `<RasterSymbolizer>` element, which contains our breakpoints. The breakpoints are defined with `<ColorMapEntry>` elements. As these elements do not have a closing tag (they use self-closing tags), we have to apply a slightly different syntax. It needs at least two attributes: a color and a raster value. The fully defined color map entry is, therefore, `<ColorMapEntry color="#000000" quantity="0"/>`. The final `<ColorMap>` element should look like the following:

```
<ColorMap>
 <ColorMapEntry color="#000000" quantity="80"/>
 <ColorMapEntry color="#FFFFFF" quantity="659"/>
</ColorMap>
```

Now we have a perfectly fine grayscale DEM. How can we use it aesthetically though? In QGIS, we applied an on-the-fly hillshading, and blended it in our land use layer. Unfortunately, GeoServer is not capable of using a hillshading effect, although we can try blending it in our land use layer. Blending is not part of the SLD specification, but available in GeoServer through a vendor option. Vendor options are parameters, which are not parts of the standard, but implemented in a spatial server to grant additional capabilities. Vendor options are defined with `<VendorOption>` elements, and are GeoServer-specific, therefore, not portable between different OWS servers.

6. The blending vendor option looks like `<VendorOption name="composite">operation</VendorOption>`, where `operation` can be any valid blending mode GeoServer knows. For our use case, the `overlay` operation would give correct results. Add the following line outside of the `<Rule>` element, but inside the `<FeatureTypeStyle>` element:

```
<VendorOption name="composite">overlay</VendorOption>
```

You can take a look at the available blending modes with some explanation at http://docs.geoserver.org/stable/en/user/styling/sld/extensions/composite-blend/modes.html.

7. Save the style, if it is valid, and apply it to the SRTM layer as a default style.

If we look at our group layer, we can see that blending modes in GeoServer work differently than in QGIS. The overlay mode keeps every cell from the DEM where there is a white background. Unfortunately, in GeoServer, we have to choose between alpha blending and color blending. With alpha blending, we can show our DEM only where we have land use polygons, but we lose their styling. With color blending, we cannot exclude the DEM where there aren't any land use polygons. To make the composition a little bit more appealing, we can constrain the DEM layer with its blending mode to a maximum scale. As a result, we will only see it on higher zoom levels, where its behavior won't distract us.

8. Go back to the style editor. We can define a maximum scale with the `<MaxScaleDenominator>` element, which should go into the `<Rule>` element. The value of the maximum scale should depend on the size of the study area. By browsing the preview map, its **Scale** should give a hint on the correct value:

```
<MaxScaleDenominator>100000</MaxScaleDenominator>
```

9. Save the edits, and preview our layer group. We shouldn't see the DEM in the initial view, while it should appear blended into the land use polygons when we zoom in enough:

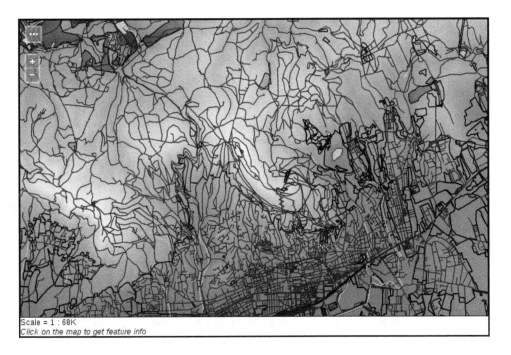

To get a composition similar to the one we created in QGIS, you have to generate a relief raster from the SRTM, clip it to the land use layer, and load the result into GeoServer (*Appendix 1.17*). Furthermore, on-the-fly hillshading is part of the SLD specification, and accessible with the `<ShadedRelief>` element. Once it is implemented into GeoTools, the processing library that GeoServer uses, it will be available in GeoServer.

Using CSS in GeoServer

We have seen that SLD has very powerful styling capabilities, although due to XML's verbose nature, SLD documents can easily become unmanageable with the style's complexity. XML is great for computers; they can easily parse and serialize documents in this format. However, they are inconvenient for humans. There is another styling language created for defining styles in a more concise manner--CSS (Cascading Style Sheet). CSS fits human logic better; we just need to understand its cascading behavior. Cascading, in this sense, can be understood with this rule--every matching definition gets applied, but on collision, the most specific definition wins. Let's look at an example. With the following snippet, we can style our GeoNames layer:

```
* {
  mark-size: 6;
}
[featurecode LIKE 'PPLA%'] {
  mark: symbol('circle');
}
[featurecode LIKE 'PPLA%'] [population > 100000] {
  mark-size: 8;
}
```

The least specific rule is the * wildcard, which selects every feature. According to that style, every feature should have a symbolizer size of 6 pixels. According to the second rule, only seats of administrative divisions should be visualized--with circles. The third rule is the most specific: it defines that seats with a population greater than 100,000 people should have an increased symbol size. What do we get in the end? Only seats are shown with a circle symbolizer. They are rendered with a 6 pixels size. Seats with a population greater than 100,000 people are rendered with an 8 pixels circle size. As the third selector is more specific than the first one, `mark-size: 8` simply overrides `mark-size: 6`.

CSS was created for styling web content in an intuitive manner. It is strongly standardized, therefore, it has not only a standard syntax, but also a standard set of values. There are some CSS variations out there, though. For example, for cartographic purposes, there is Mapbox's

CartoCSS, Mapnik's Cascadenik, and GeoServer's CSS extension. They share the same syntax, although their set of values are completely vendor specific. They are unrelated to basic CSS, and in some cases, even to each other.

The syntax of CSS is very intuitive once we get a grasp of it. Stylesheets written in CSS consist of rule blocks. Each block defines a single set of rules the renderer should apply on logically coherent items (features or raster cells in our case). Every block has two parts--a **selector** and a **declaration block**. A declaration block (in braces) has individual declarations as key-value pairs separated by a colon, and terminated by a semicolon. The most basic selector is `*`, which selects everything. It is like creating a `<Rule>` element in SLD without any filters. In square brackets, we can define filter expressions. We can put multiple selectors together to form a more specific selector. If we separate multiple selectors with whitespaces, we form a logical `AND`, while if we separate them with commas, we form a logical `OR` operation between them.

 As in QGIS, you can use the keywords `AND` and `OR` in selector expressions. GeoServer's CSS processor still treats the selector as more specific than a single value query. For example, `[featurecode LIKE 'PPLA%']` `[population > 100000]` has the same specificity as `[featurecode LIKE 'PPLA%' AND population > 100000]`.

As GeoServer's renderer is built around the SLD specification, its CSS also has SLD capabilities--just with another, more convenient syntax. That is, we can use SLD rules without nesting elements into each other, keeping only the important parts of styling. Of course, as the language is different, there are also changes in the syntax.

 Fortunately, as CSS styling is an official extension, it has documentation accessible from GeoServer's official site at `http://docs.geoserver.org/l atest/en/user/styling/css/index.html`. From there, you can read out the attributes, and other coding styles required to use GeoServer's CSS.

In order to have CSS capabilities in GeoServer, we first have to download and install the extension. As it is an official extension, we can download it from GeoServer's download page at `http://geoserver.org/release/stable/`:

1. Download the **CSS Styling** extension at the bottom of the page.
2. After downloading and opening the archive, extract its content into the GeoServer's library folder. It can be found in the GeoServer's folder under `webapps/geoserver/WEB-INF/lib`.
3. Restart GeoServer to load the extension.

Styling layers with CSS

First, let's style the land use polygons. It would be very cumbersome to set up all the filters in SLD, however, in CSS, we have an easy job creating different categories.

1. Create a new style. Name it accordingly, and limit it to our workspace.
2. Select the **CSS** option in the **Format** field.
3. Generate a polygon template, save the style, reopen it, and preview it on our `landuse` layer.
4. The template contains a fill style, a stroke style, and a comment line with a `@title` directive. As we have no means to name legend labels in CSS, when GeoServer converts the stylesheet to SLD, it extracts label titles from specially formatted comment lines containing only a `@title` directive and a title name. Rewrite the block to show only forests with only a fill as follows:

```
/* @title Forest */
[fclass = 'forest'] {
 fill: #1ea914;
 }
```

5. Finish the style by defining the rest of the declaration blocks with the appropriate selectors. When we would like to apply a single style to multiple classes, we can use multiple selectors separated by commas:

```
/* @title Grassland */
[fclass = 'farm'], [fclass = 'grass'], [fclass = 'meadow'],
 [fclass = 'vineyard'], [fclass = 'allotments'] {
  fill: #b2df8a;
 }
 /* @title Residential */
 [fclass = 'residential'] {
   fill: #fdbf6f;
 }
 /* @title Industrial */
 [fclass = 'industrial'], [fclass = 'quarry'] {
   fill: #b3b3b3;
 }
```

6. If the stylesheet is valid, save it, and make it the default style of the `landuse` layer:

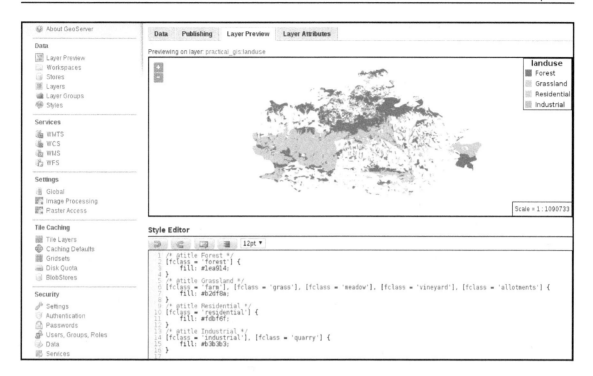

As the style definitions are mutually exclusive (there are no common rules), GeoServer has an easy job converting the title directives to legend labels. However, if we use some overlapping rules, and they have titles, GeoServer will concatenate the different titles into a single label. Currently, there is no workaround to solve that issue.

Creating complex styles

Next, let's create some complex styling to visualize our main roads similar to the visualization we created in QGIS:

1. Create a new style. Name it accordingly, and limit it to our workspace.
2. Select the **CSS** option in the **Format** field.
3. Generate a line template, save the style, reopen it, and preview it on our `roads` layer.
4. Rewrite the rule as follows to show secondary roads, as they had a simple symbolizer:

```
/* @title Important roads */
[fclass LIKE 'secondary%'] {
```

```
  stroke: #8f9593;
  stroke-width: 1;
}
```

5. It's time to create the complex line styles. In SLD, we would have to create multiple <FeatureTypeStyle> elements to have multiple lines drawn on each other. In CSS, however, we can define multiple styles in a single definition separated by commas. Similarly, we can define their other attributes the same way. The order of the values is the only thing what matters:

```
/* @title Motorways */
[fclass LIKE 'motorway%'] {
  stroke: #000000, #eff21e;
  stroke-width: 4, 3;
}
/* @title Highways */
[fclass LIKE 'primary%'] {
  stroke: #000000, #ded228;
  stroke-width: 3, 2;
}
```

Note that in QGIS, we defined line widths in millimeters. In GeoServer, we usually define widths in pixels. Therefore, you might need to fiddle with the width values a little bit to get aesthetic results.

Our complex line styles should look like the following:

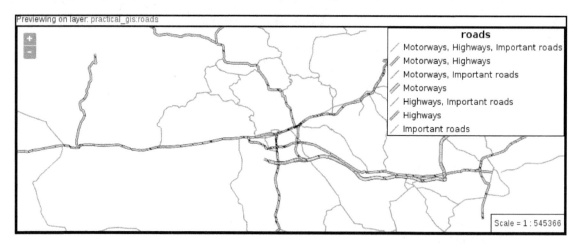

6. We've got some better results, although our map still suffers from the symbol levels problem like in QGIS. In GeoServer, for altering the order of rendering lines with different styles, we can use the `z-index` property. When we have multiple styles assigned to a single-line type, we have to use multiple z-indices separated by commas. Before applying the symbol levels, find out the correct order. The black borders should come first, while the rest of the roads should be drawn according their priorities. Important roads second, highways third, and motorways, fourth:

```
/* @title Motorways */
[fclass LIKE 'motorway%'] {
  stroke: #000000, #eff21e;
  stroke-width: 4, 3;
  z-index: 1, 4;
}
/* @title Highways */
[fclass LIKE 'primary%'] {
  stroke: #000000, #ded228;
  stroke-width: 3, 2;
  z-index: 1, 3;
}
/* @title Important roads */
[fclass LIKE 'secondary%'] {
  stroke: #8f9593;
  stroke-width: 1;
  z-index: 2;
}
```

You can see the available CSS properties in the official reference at http ://docs.geoserver.org/latest/en/user/styling/css/properties.ht ml.

Now we should be able to see clean lines rendered by GeoServer:

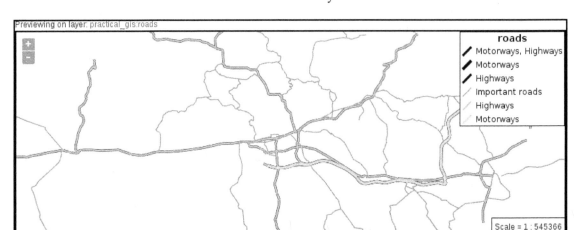

7. The only problem left is the occasional gaps between our line segments. As GeoServer applies a butt line ending by default (*Appendix 1.18*), some of our segments get cut off in their meeting points. To solve this issue, we can round off all of our lines with a global definition as follows:

```
* {
  stroke-linecap: round;
}
```

It is a good practice to use clean, processed, visualization-ready vector data in GeoServer. Although GeoServer has some quite advanced capabilities to handle raw vector layers, it is far from a full-fledged GIS software.

8. Finally, let's add some labels showing road references to important roads. If we add labels globally, we would end up with a map showing labels for every road no matter whether they are visualized or not. Therefore, we need a selector which selects only our features of interest. The labels should have a white color, and a bluish rectangular background. We can set a background with the shield property, which accepts a point symbolizer as a value. As we have a square symbol at hand, we can use that. When we put the code together, we should get something like the following:

```
[fclass LIKE 'motorway%'], [fclass LIKE 'primary%'],
  [fclass LIKE 'secondary%'] {
  label: [ref];
  font-family: DejaVu Sans;
```

```
font-fill: #ffffff;
shield: symbol('square');
}
```

9. Although we have labels on our map, it still bleeds from several wounds. First of all, we need to offset our labels so that they are placed in the middle of our lines. We can achieve this by modifying the anchor point. The anchor point defines the reference point of our labels. It is placed in the middle of our lines, and the label is drawn from that point. By default, the anchor point is the lower-left point of our labels, which is represented with two 0s. As the upper-right coordinates are represented with two 1s, the middle point of our labels are two 0.5s:

    ```
    label-anchor: 0.5 0.5;
    ```

10. We should also remove duplicated labels. The default behavior of GeoServer is to render a label on every separate segment. As we have many segments, we get a lot of labels. Although there is no SLD option for merging logically coherent lines, GeoServer can do that with a vendor option. In CSS, we can also use vendor options prefixed with `-gt-`. The correct option for this is `-gt-label-group`, which renders a single label for lines with the same label attribute, on the longest segment:

    ```
    -gt-label-group: true;
    ```

11. Now we have fewer labels; however, regardless of their width, they are rendered on the same-sized square. We can override this behavior by using two vendor options---`gt-shield-resize` and `-gt-shield-margin`. The former defines how GeoServer should resize shields when the label sticks out, while the latter defines the margin size around the labels in pixels:

    ```
    -gt-shield-resize: stretch;
    -gt-shield-margin: 2;
    ```

12. The only thing left to do is to customize the shields. Markers can be customized by using pseudo selectors to apply further properties to every symbol. In this case, we can safely customize every shield marker by using a sole `:shield` pseudo selector in a separate definition block, as we have only one type of shield.

```
:shield {
  fill: #244e6d;
}
```

Putting the whole code together, we get a label description similar to the following:

```
[fclass LIKE 'motorway%'], [fclass LIKE 'primary%'],
[fclass LIKE 'secondary%'] {
label: [ref];
font-family: DejaVu Sans;
font-fill: #ffffff;
label-anchor: 0.5 0.5;
shield: symbol('square');
-gt-label-group: true;
-gt-shield-resize: stretch;
-gt-shield-margin: 2;
}
:shield {
  fill: #244e6d;
}
```

 If you have multiple types of shields to style individually, you can nest the definition block of the shield pseudo element into the definition block of the labels using it. You can read more about nested rules at `http://docs.geoserver.org/latest/en/user/styling/css/nested.html`.

If we apply the final style as the default on the roads layer, and override the road layer's style of our layer group in **Data | Layer Groups**, we can see our final composition in **Data | Layer Preview**:

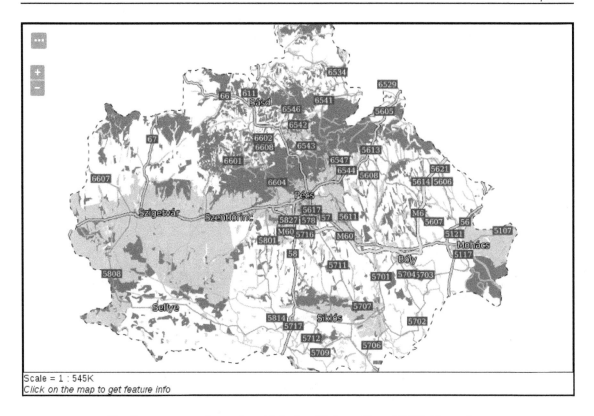

Scale = 1 : 545K
Click on the map to get feature info

GeoServer cares less about label collisions than QGIS. To reduce overlapping labels, you can increase the minimum required space in pixels between adjacent ones with the `-gt-label-padding` vendor option.

Styling raster layers

As a final task, let's style our suitability raster layer. We should use the same color ramp that QGIS calculated for us. That is, we have five breakpoints: `0`, `0.25`, `0.5`, `0.75`, and `1`. The color ramp is red to green, with the colors `#ff0000`, `#fdae61`, `#ffffc0`, `#a6d96a`, and `#1a9641` at the breakpoints:

1. Create a new CSS style restricted to our workspace with an automatically generated raster template. Save the style, reopen it, and preview it on our suitability layer.

2. Single-band rasters can be styled in CSS just like in SLD--with a color map. The syntax of the color map is a bit unusual, though. The property is `raster-color-map`, while the value is a set of `color-map-entry` functions separated by whitespaces. As a one liner would be hardly manageable for this rule, and CSS allows some flexibility in its syntax, we can put our color map entries in separate lines. A single `color-map-entry` function accepts three arguments: a color, a value, and an optional opacity value. Extend the definition block with a color map as follows:

```
raster-color-map:
  color-map-entry(#ff0000, 0)
  color-map-entry(#fdae61, 0.25)
  color-map-entry(#ffffc0, 0.5)
  color-map-entry(#a6d96a, 0.75)
  color-map-entry(#1a9641, 1);
```

3. The result is correct, although we get a lot of red pixels due to the amount of unsuitable areas. To get rid of zero values, we can provide a 0 opacity value in the first color map entry, making red areas fully transparent:

```
color-map-entry(#ff0000, 0, 0)
```

The red parts are gone, however, GeoServer not only interpolates colors between breakpoints, but also opacity values. That is, the opacity between 0 and 0.25 is constantly changing, introducing a dull reddish color in that interval. To make things worse, layers beneath the suitability layer will be visible in those areas, distorting the colors even more. The logical solution would be to increase the value of the lowest color map entry by an arbitrarily low value (like 0.00001). The problem is, GeoServer not only interpolates on a color ramp, but also extrapolates. That is, if we do not set zero values to fully transparent, we always get them styled according to the color of the lowest color map entry. The only thing we can do is to introduce another entry, which jumps the opacity of the layer back to normal on values slightly bigger than 0.

4. Insert an additional color map entry with a red color (#ff0000), a very low value (like 0.00001), and an opacity value of 1:

```
color-map-entry(#ff0000, 0.00001, 1)
```

5. Save the style, and assign it to our suitability layer as default:

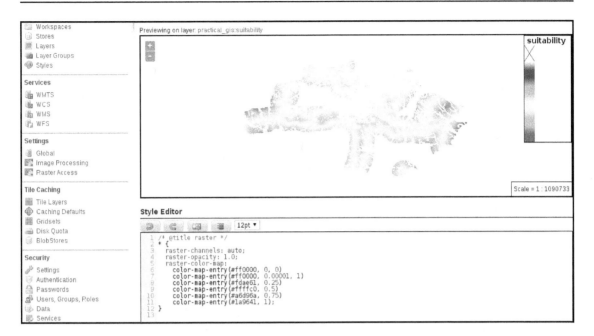

You can also interpolate colors with vector layers to create thematic maps, such as a choropleth. To learn more about these transformation functions, you can read the official documentation at `http://docs.geoserver.org/latest/en/user/styling/css/examples/transformation.html`.

Summary

In this chapter, we learned the basics of styling data in GeoServer. We discussed the symbolizers we can use, how XML and the XML-based SLD styling language work, and finally, how we can create styles easily with CSS. You were introduced to the syntax of the aforementioned styling languages, and now can create styles for our vector and raster maps from simple visualization purposes to more advanced cartographic use cases.

In the next chapter, we will learn about the client side. We will explain how client-side web mapping software works, and how we can utilize them to request spatial data from the server's file system or a spatial server. We will also see how we can use spatial data with JavaScript, and how we can script a web mapping software to create interactive maps with already-styled images and raw vector data.

13
Creating a Web Map

In the previous chapter, we learned how to style spatial data in GeoServer. We started with a simple symbology for raw vector and raster data and proceeded to more advanced and also more aesthetic cartographic representations. We ended with a group layer resembling the composition we created with QGIS's print composer. We also looked at some other independent vector and raster layers, styled and ready for use.

Now we will use our styled data and create some client-side interactive maps to showcase our results. We will use JavaScript to create a web map that can not only use our already styled vector and raster data as image layers using WMS, but can also use raw vector data with WFS. We will cover how to make our maps more interactive by styling vector data on the client side and enabling our users to query them without sending additional requests to the server.

In this chapter, we will cover the following topics:

- The basic JavaScript syntax
- Using the Leaflet API
- Displaying styled maps as images
- Using raw vector data

Understanding the client side of the Web

In Chapter 11, *Showcasing Your Data*, we discussed how data is transferred over the Web and how servers work. In order to have a better understanding of the Web, let's discuss how web clients interpret server responses in more detail. As we already know, servers either store web content in a static format, or they generate it on the fly with CGI scripts or other web applications.

We also know that these contents are usually plain text, structured text, or media files. The most common content a web client has to interpret is in structured text format, containing elements we would like to show, styles we would like to apply to our elements, and scripts we would like to run on the client side:

- **HTML**: Hypertext Markup Language is the standardized form of transferring visual elements from web servers to web clients. They are XML-based documents that describe each visual element between tags. Although HTML is XML-based, a valid HTML document is not necessarily a valid XML document. For example, the HTML standard does not make self-closing single tags mandatory. If we write `<br>` in a HTML document, it is a valid HTML; however, we have to write `<br/>` to get a valid HTML and XML document.
- **CSS**: Cascading Style Sheets is the standardized way to describe the custom styling of HTML elements. Every web client has a default set of styling options that are applied to HTML elements with no custom styles. If the web client gets custom rules in the form of CSS declaration blocks, it overrides the default styling with them.
- **JavaScript**: We can also use custom scripts written in JavaScript in order to send executable code to the client. The client interprets and runs the code contained in the JavaScript file, enhancing the user experience by making the web page more dynamic. It is very useful to automate smaller tasks without wasting the server's resources. For example, interactive web maps are created with web mapping libraries. Web mapping libraries are essentially collections of JavaScript functions creating interactive maps based on some parameters we provide.

What happens to these documents in the web client? First of all, the client sends a request to the destination URL. If we did not provide a resource name and just a path, the request will default to the `index.html` document in the provided path. Then, if a web server listens on the other side, the communication gets established, and the transaction we discussed earlier occurs. The client receives a response, which is some kind of resource (most often an HTML document). If the HTML document contains links to other resources (for example, stylesheets, scripts, and media elements), the client requests these items individually and interprets their content. If a stylesheet is requested, the client applies the styles found in there, while if a script is requested, it parses and executes it.

 Modern web browsers are smart. They try to get the most out of the received data. For example, they can open raw PDF or media files and automatically generate a DOM model when there is a raw file in a recognized format on the other side of the connection.

Let's assume the resource is an HTML document. The client parses the elements written in HTML and creates an object model from it, called a **DOM** (**Document Object Model**). The DOM is the object-oriented representation of the HTML document using a tree structure. Every element is an object, with the various attributes the element can hold. We can interact with the DOM, query, modify, insert, and remove individual elements in it. Of course, we need a way to interact with the DOM. As web clients expose their DOM trees through their JavaScript interfaces, we can manipulate DOM elements through JavaScript. For example we can query input values and act accordingly. To make this interaction more convenient, the JavaScript DOM API comes with a built-in event model, that is, we can register event listeners on DOM elements, and the registered functions get executed automatically every time the event occurs.

Creating a web page

To understand the client side better, let's make a simple web page containing some basic information about our map. First of all, we need to get our web server's root directory. As web servers can only see a portion of the filesystem, they can only serve our documents if they are placed in the portion they can use. Apache comes with a default web page, which can help us locate this root folder without searching for configuration files. If we open `http://localhost` in a browser, we can see the greeting document we are searching for. The document's name in the filesystem is `index` with a varying extension (for example, `html`, `html.en`, and `phtml`).

> Note that on Windows you have to start Apache manually every time you start your system, if you're using the OSGeo4W version of it.

The location of the web server's root folder can vary between different operating systems, versions, and distributions.

- On most Linux distributions, the web server's root folder is located somewhere in `/var/www`. On Red-Hat-based distributions (for example, Fedora, CentOS), it is in `/var/www/httpd`.
- On macOS systems, it is either located in `/Library/WebServer/Documents` or `/usr/htdocs`.
- On Windows, it is located in `C:\OSGeo4W\apache\htdocs`, assuming Apache was installed with the OSGeo4W installer and the default path.

If none of these paths work, you can search for the Apache configuration containing Apache's root folder on your filesystem, which is called either `000-default.conf`, `apache.conf`, or `httpd.conf` depending on your OS and Apache version. The line you should be looking for looks like `DocumentRoot /var/www`, where instead of `/var/www`, you will see your Apache root folder's absolute path. If nothing helps, a Google search with your OS's name and version can also help.

After we locate the root folder of our Apache web server, we are only a few steps away from creating our first web page:

1. Make sure you have write permission to Apache's root folder.
2. Create a new file called `map.html`.
3. Open the file with a text or a code editor. A good code editor with syntax highlighting can gradually help correct typing errors. More advanced editors (for example, Atom, Visual Studio Code) can even spot some syntax errors.

Writing HTML code

HTML has a syntax similar to XML and SLD. The basic principles are the same; we use opening and closing tags, tag attributes, and content. The content can be plain text or other HTML elements. There are only a few elements that we will use to create our web map:

- **<html>**: The root element of every HTML document grouping the whole content.
- **<head>**: The group of important elements that are not visualized but alter the default behavior of the web page in some way.
- **<body>**: The group of visual elements rendered by the web client.
- **<title>**: The title of the page. It goes inside the `<head>` element.
- **<meta>**: The various kinds of metadata of the web page. Very useful to explicitly set the character encoding of the web page.
- **<link>**: A reference to an external resource, usually a stylesheet.
- **<script>**: Inline JavaScript code or a reference to an external resource containing valid JavaScript code.

- **<h1>**: A first-level header emphasized with a large font size.
- **<p>**: A paragraph of plain text.
- **<div>**: A division without much purpose on its own. When it contains other elements, it groups them together. This way, visually coherent parts of the website (e.g. navigation bar, main content, sidebars) can be grouped and styled easily with CSS. It is used by web mapping libraries to group map content.
- **<input>**: The various user inputs based on the attributes of the tag. It can be a radio button, a checkbox, a regular button, a text input, or a numeric input, just to mention a few.

 You can look at the complete HTML tag reference at `https://developer.mozilla.org/en/docs/Web/HTML/Element`. When you need some help with frontend development, **MDN (Mozilla Developer Network)** is one of the most trustworthy sites to gather information.

There are some basic rules and guidelines to create a good HTML document. Every element should go in the `<html>` root element. The root element usually has a single attribute (for example, `lang="en"`), containing the language abbreviation of the page. Furthermore, it should only contain the `<head>` and `<body>` elements. The `<head>` element should contain everything that is not for direct visualization. Elements such as `<title>`, `<link>`, and `<meta>` should all go into the `<head>` section. Additionally, JavaScript libraries needed for the web page's custom code should also go into the `<head>` section. The `<body>` section should contain the rest of the elements, that is, everything that should be rendered. It can also contain the web page's custom scripts, especially if they are changing the other elements; therefore, they require the DOM tree to be set up by the client when they run. A minimal functional example of the aforementioned elements looks like the following:

```
<html lang="en">
  <head>
    <title>Your page's title</title>
    <meta charset="utf-8">
    <link href="yourstylesheet.css" rel="stylesheet">
    <script src="yourscript.js" type="text/javascript">
  </head>
  <body>
    The content of your web page.
  </body>
</html>
```

 It is more elegant to register the custom code to the `<body>` element's ready event and include it in the `<head>` section.

Apart from the custom attributes specific elements can hold, we should be able to apply some general attributes, mostly on visual elements. With the `id` attribute, we can assign a unique identifier to individual elements. Using IDs, we can style these elements individually and also get their object references through JavaScript. This is the way in which we can register listener functions to elements, modify them, or query their attributes in our custom code. With the `class` attribute, we can group elements. Using classes is mostly used to style visually coherent elements with a single CSS rule.

Now that we know about the basic principles, let's create our HTML document. The web page should have a title and a UTF-8 character encoding in its `<head>` section. In the `<body>` section, we should have a header with a greeting text and two `<div>` elements. The first one will contain the map, while the second one will contain a short description. As we would like to be able to select between multiple maps, we should have at least four buttons under the `<div>` elements. The four maps we will create are the road map, the constrained houses from our PostGIS analysis, the suitability map, and the road map in our local projection.

After putting the required elements together, we should get a simple HTML document similar to the following:

```html
<html lang="en">
 <head>
   <title>Map gallery</title>
   <meta charset="utf-8">
 </head>
 <body>
   <h1>Welcome to my interactive map gallery!</h1>
   <div id="map"></div>
   <div id="description">
     <p id="description_text"></p>
   </div>
   <input id="roadmap" type="button" class="mapchooser"
    value="Road map">
   <input id="suitability" type="button" class="mapchooser"
    value="Suitability">
   <input id="houses" type="button" class="mapchooser"
    value="Houses">
   <input id="roadmap_23700" type="button" class="mapchooser"
    value="Road map (EPSG:23700)">
 </body>
</html>
```

You can read about the appropriate attributes of an HTML element at the corresponding MDN site. For example, the `<meta>` tag's description can be reached at `https://developer.mozilla.org/en/docs/Web/HTML/Element/meta`.

If we save our edits made to our HTML document and open it in our browser (`http://localhost/map.html`), we will able to see our raw HTML elements rendered with their default styles:

Styling the elements

Although the syntax is the same, regular CSS differs from GeoServer's CSS. We already know that CSS rules consist of selectors and declaration blocks, which contain declarations in the form of key-value pairs. There are three kinds of basic selectors we should be able to use in CSS. We can style every tag in an HTML document by supplying the tag's name in a selector. For example, a `div` selector selects every `<div>` element in the document. We can constrict the selection to a specific class using `.classname` and to a specific ID using `#idname`. Selection based on multiple criteria (the AND logical operator) can be achieved by writing the corresponding selectors together. By writing `div.map`, we select every `<div>` element that has a `map` class. We can also define a union (the OR logical operator) by separating different selectors with commas. Writing `div, .map` selects every element that is either a `<div>` element or has a `map` class. Finally, by separating different selectors with only whitespaces, we can look for parent-children relations. By writing `div .map`, we select every element with a `map` class, which is directly nested into a `<div>` element. There are also special pseudo selectors that check for specific events. For example, if we use a `:hover` pseudo selector with a `div` selector and write `div:hover`, the declarations are only applied when we are hovering over a `<div>` element with our cursor.

The declarations are also standardized in regular CSS. You can see a list of the available declarations at `https://developer.mozilla.org/en-US/docs/Web/CSS/Reference`. Don't worry about invalid declarations. They don't raise an error; they simply get discarded by the web client. This makes plenty of room for vendor-specific options in different web clients.

Let's learn some basic styling tricks by styling our document:

1. Add some multiline placeholder text to our `<p>` element (`<p>text</p>`).
2. Create a new file in our web server's root folder called `map.css`.
3. Link this CSS file to our HTML document so that we can see our changes instantly. Use a `<link>` element in the `<head>` section to create this reference. The element must have two attributes: a `href` attribute pointing to the CSS file and a `rel` attribute containing the `stylesheet` value. The `<head>` section should look like the following:

```
<head>
 <title>Map gallery</title>
 <meta charset="utf-8">
 <link href="map.css" rel="stylesheet">
</head>
```

Although it is out of the scope of this book, learning to use the client's developer tools is an invaluable asset in frontend development. You can reach the developer tools in most of the modern browsers by pressing *F12*, *CTRL* + *SHIFT* + *I*, or *CTRL* + *SHIFT* + *J*.

4. Open `map.css` with a text or code editor.
5. Create a declaration block for our `<div>` elements. For now, declare only a `height` value. We can supply physical sizes with absolute and relative values in CSS. An absolute value is usually supplied in pixels (for example, `20px`), while a relative value is often supplied as a percentage value (for example, `80%`). Relative values are calculated by the web client on rendering. They are always relative to the parent element's size:

```
div {
 height: 80%;
}
```

6. Create two declaration blocks--one for the map element and one for the description element. Use the IDs of the elements in the selectors. Give the map element a width of `60%` and the description element a width of `40%`:

```
#map {
  width: 60%;
}

#description {
  width: 40%;
}
```

7. Now we have our two `<div>` elements sized properly, but they aren't placed on the same line. The problem is that they are block elements. Block elements can be sized arbitrarily, but they are always placed under each other. To override this behavior, we have to float our map element to the left and declare both of the elements as inline blocks. Inline blocks can be sized arbitrarily, but they can also be floated next to each other until the sum of their sizes does not exceed the viewport's size:

```
div {
  height: 80%;
  display: inline-block;
}

#map {
  width: 60%;
  float: left;
}
```

When designing web pages, it is a good practice to check the result once in a while on Firefox. Firefox is mostly sticking with the standards, so if a design is working on Firefox, it will probably work on other browsers as well. Other browsers (especially Chrome) try to outsmart the developer sometimes, making style combinations work, which shouldn't happen according to the standards.

8. Add a thin border around the map element to have a nice placeholder, even if we do not have a loaded map. Adding a border is one of the few things for which CSS offers a convenience property. It is called `border`, and we can provide the three most important properties of a border as values: its width, its color, and its line style. The order is not important; however, the values must be separated with whitespaces. The rule should look like the following:

```
border: 1px solid black;
```

9. Now if we refresh our site, we can notice that the design is broken and the elements are placed under each other again. The problem is that the borders, padding, and margins are not included in the overall size of an element by default. That is, we have exceeded the width of the viewport by two pixels. We can fix that by overriding this behavior for the map element. We can do that with an extra rule--`box-sizing: border-box`, although it only applies to borders. Padding and margins are still not included in the specified size. We can also round down the corners of the border nicely by supplying a `border-radius` property with an arbitrary value:

```
#map {
  width: 60%;
  float: left;
  border: 1px solid black;
  border-radius: 5px;
  box-sizing: border-box;
}
```

10. The only thing left to do is increase the space between the map and the other elements (text and buttons). As defining a margin or a padding on either of the `<div>` elements would result in a broken design again, we can define a padding on the `<p>` element in order to achieve the desired effect. By adding additional rules to the `<div>` elements, we can safely create a padding:

```
div {
  height: 80%;
  display: inline-block;
  padding-bottom: 1em;
}

p {
  padding-left: 1em;
}
```

The em unit is another frequently used relative length. It is relative to the size of the styled element's currently used font. Using relative sizes is one of the basic principles of responsive web design.

Our basic map gallery should look similar to the following:

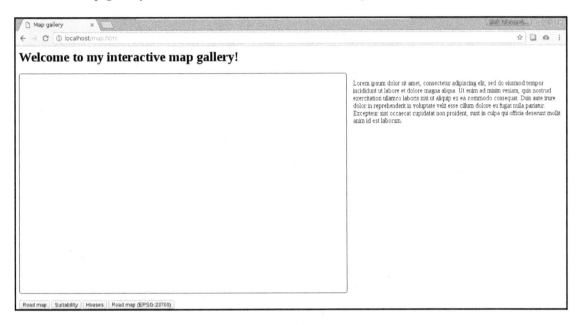

Finally, let's style our buttons. We can go wild here without worrying about breaking the page design:

1. Create a declaration with a selector, selecting the class used for our buttons.
2. Create a declaration block with some properties of your taste. A very simplistic solution can be the following one:

```
.mapchooser {
height: 40px;
border: 1px solid black;
border-radius: 5px;
background-color: white;
}
```

3. If we inspect the result, we can see our new button design, although no visual changes occur when we click on them. As we overrode the default button styling, we also have to supply some rules for the click event. Define a style for the click event using the `:active` pseudo selector:

```
.mapchooser:active {
  border: 2px solid black;
  background-color: grey;
  color: white;
}
```

We can use as many CSS rules as we would like on our buttons, although using the simple rules provided here still brings a significant change in the design of our buttons:

Scripting your web page

The third component of a web page is the custom code we script our page with. We can make our page more dynamic and interactive by executing different tasks on different events, for example, on a press of a button. Web clients use the JavaScript language for these kinds of client-side tasks. They expose their DOM trees, making them accessible and modifiable by us. JavaScript started as a high-level basic scripting language for very simple automation tasks. Nowadays, it can be considered a full-fledged high-level object-oriented programming language. DOM manipulation is just one of the many things it can do, including the interpretation and manipulation of binary input. It can be used in imperative style, object-oriented style, and--to some extent--functional style. For the sake of simplicity, we will stick with a minimal subset of its capabilities and write simple yet effective procedural code.

Let's discuss the syntax of JavaScript briefly. In the code we will use, we will encounter some basic types that behave differently, such as literals:

- **Integer**: Whole numbers without decimal places. They can be represented with numeric characters (for example, 1).

- **Floating point**: These are decimal numbers. They can be represented with numeric characters, with the point character as the decimal point (for example, `0.25`).
- **String**: Character strings consisting of a sequence of characters. They must be enclosed in quotation marks. They can be single or double quotation marks; however, both the opening and closing mark must be of the same type (for example, `"character"` or `'string'`). They can contain only numeric characters but are still considered character strings.
- **Boolean**: A binary value. It can be either `true` or `false`.
- **Array**: A set of other literals separated by commas, enclosed in brackets (for example, `[0, 1, 2]`). Arrays can contain mixed types, although in most cases, it is not a good practice.
- **Object**: A set of key-value pairs separated by commas, enclosed in braces. The keys and the values are separated by colons (for example, `{x: 5, y: 10}`). Keys must be representable by strings, while values can be of any type. Keys do not have to be enclosed in quotation marks if their conversion to strings is simple, although the grammar of the values must adhere to the grammar of the represented types.

> You can read more about JavaScript grammar at `https://developer.moz illa.org/en-US/docs/Web/JavaScript/Guide/Grammar_and_types`.

A non-literal type we should also be able to use is the function. Functions group individual statements (commands) and execute them sequentially. We can call a function by its name and some parameters. The number of parameters a function can accept always depends on the given function (functions accept zero or more parameters). The correct syntax calls the function's name and places the different parameters after it in parentheses, separated by commas (for example, `myFunction(x, y, z)`). There must be no whitespaces between the function's name and the opening parenthesis:

```
// We have a function named addNums requiring two numbers
   and returning the sum of them.
addNums(2, 5);
```

> Functions are often called methods, especially if they are in objects, as objects can store functions as values. The two terms have some differences, although most of the time, they are used interchangeably.

We should also know how to use variables. Variables can be declared in multiple ways, although using the `var` keyword fits our needs. A variable declaration needs the `var` keyword, a variable name, and, optionally, an assignment with an initial value:

```
var x; //Declaring a variable named x.
var x = 2; //Declaring a variable named x initialized with a
 value of 2.
```

 TIP

Single-line comments can be used with the `//` sequence. Everything after `//` will be considered a comment and won't be evaluated. Multiline comments can be opened with the `/*` sequence and closed with the `*/` sequence.

Accessing variables can be done by simply providing the name of the variable. Different types act differently, though. Arrays and objects can contain multiple elements, which can be accessed directly. Individual array elements can be accessed by their index numbers with the subscript operator (brackets) starting with the index number 0, while object elements can be accessed by separating the object's name and the key's name with the dot operator (point):

```
var x = 2;
var array = [9, 4, 7];
var object = {x: 5};

x; //2
array[1]; //4
object.x; //5
```

Finally, most of the statements must be terminated. There are some exceptions, such as control flow statements (for example, loops and conditionals); however, in the case of declarations, assignments, and function calls, it is mandatory. The symbol we should terminate our statements with is the semicolon. Although statements can be terminated with line breaks instead of semicolons, and the code will run in most cases, it is a very bad practice, as it makes the code vulnerable. For example, if we want to compress our script without using semicolons, the whole code will break.

For now, let's create a script that shows the description of a map when we click on the button representing it. First of all, we just declare a variable containing a single description and show it on our web page:

1. Create a new file in the web server's root folder, named `map.js`.

2. Edit the `map.html` file and include the JavaScript file in a script tag. As our code will require the DOM to be set up, we should include it in a `<script>` element at the end of our `<body>` section. A script element referencing an external resource needs an `src` attribute with a relative path and a `type` attribute with the `text/javascript` value:

```
<body>
[...]
<input id="roadmap_23700" type="button" class="mapchooser"
value="Road map (EPSG:23700)">
<script src="map.js" type="text/javascript"></script>
</body>
```

3. Edit the `map.js` file with a text or code editor.

4. Declare a variable with a string describing our road map:

```
var roadmap = 'Road map of Baranya county. Land use types,
    important roads, major settlements, and basic hydrology
    are visualized.';
```

5. The only thing left to do is update our web page with this string. We can access the root of the DOM tree with the `document` variable. It is an object that contains various methods to access individual DOM elements. A very popular method is `getElementById`, which simply requires the unique ID of the queried element and returns it. Query the `<p>` element containing the description and save it to a variable:

```
var description = document.getElementById('description_text');
```

6. Now we have our `<p>` element's DOM representation saved into the `description` variable. It has a lot of methods and attributes with different purposes, such as registering event listeners or changing the element. We need its `textContent` attribute, which contains the visualized text. By changing that attribute to the content of our `roadmap` variable, the visualized text changes as well:

```
description.textContent = roadmap;
```

By refreshing the page, we will be able to see the description of our road map next to the map container:

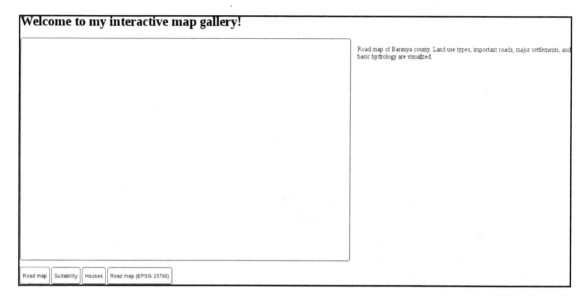

In order to register these lines to the click event of our first button, we have to make a procedure from them. Procedures in JavaScript are functions, which we already know how to call. We only have to learn how to make one. Creating a function involves the `function` keyword, some parameters our function will use in parentheses, and a code block in braces. The code block contains the statements our function will execute on call:

1. Create a working function from the three lines we created. The function is called `updateMap`, and it does not require any parameters for now. We still have to include an empty parameter list in parentheses:

    ```
    var updateMap = function() {
    var roadmap = 'Road map of Baranya county. Land use types,
    important roads, major settlements, and basic hydrology
    are visualized.';
    var description = document.getElementById('description_text');
    description.textContent = roadmap;
    };
    ```

2. If we refresh our page, the placeholder text is still there, and we can see that it did not get updated to our description automatically. Let's open the developer tools, navigate to its **Console** tab, and test our function by calling it (`updateMap();`). The text is now updated.

If you can still see the updated text instead of the placeholder when refreshing the web page, your browser might be using a cached version of the page. Try to refresh the page with the developer tools opened (*F12, CTRL + SHIFT + I*, or *CTRL + SHIFT + J*).

3. Register this new function on our first button's click event. The event model in JavaScript is quite capable, but it offers convenient methods for simple tasks. Registering only one function to a popular event (for example, clicking) is such a task. We have to get the reference to the button's DOM element and assign the function to its `onclick` attribute. The client will automatically call the function every time we click on the button:

```
var roadbutton = document.getElementById('roadmap');
roadbutton.onclick = updateMap;
```

In JavaScript, we can freely chain statements. They will be evaluated in order, and the result of a prior call will be used in the next call. Therefore, the previous lines can be written in a one-liner of `document.getElementById('roadmap').onclick = updateMap;` without saving anything in variables.

Great work! Now we have the placeholder text, and we can change it to our description with a click of a button. The only problem is that we have to repeat this process for every button we have. On the other hand, we can shape our function into a more general form that can find out the assignable text based on the ID of the clicked button. Then, we can register the same function to every button we have. The basis for our new logic is to create an object that stores the descriptions of the maps with keys representing the IDs of the corresponding buttons. In the event listener, we will read the correct description based on the clicked button's ID:

1. Create an object that stores the descriptions and save it to a variable. The keys must exactly match the IDs of the buttons. The object must be declared before the function:

```
var descriptions = {
  roadmap: 'Road map of Baranya county. Land use types,
    important roads, major settlements, and basic hydrology
    are visualized.',
  suitability: 'Suitability map of Baranya county for
    building a new warehouse for a logistic company having
    stores in major settlements.',
  houses: 'Real estates in Pecs residing in nice,
    quiet areas with shops, restaurants, bars, parks,
    and playgrounds nearby.',
```

```
roadmap_23700: 'Road map of Baranya county in EOV
  projection.
  Land use types, important roads, major settlements,
  and basic hydrology are visualized.'
};
```

2. Generalize the `updateMap` function to read out the correct description based on the clicked button's ID. For this, we need a reference to the clicked button. Luckily, the event model in web clients takes off this weight. When the client calls the listener function, it tries to pass an event parameter to it. If it can, we can access the corresponding DOM element from the event parameter's `target` attribute. The only problem left is that we don't know how to access an object's attribute with a variable. We can also access object members with another accessor. Using bracket notation, we can query an attribute with its key as a string literal or with a variable:

```
var updateMap = function(evt) {
 var description =
   document.getElementById('description_text');
 description.textContent = descriptions[evt.target.id];
};
```

3. Register our new, general function on every button we have:

```
document.getElementById('roadmap').onclick = updateMap;
document.getElementById('suitability').onclick = updateMap;
document.getElementById('houses').onclick = updateMap;
document.getElementById('roadmap_23700').onclick = updateMap;
```

4. Edit the `map.html` file and remove the placeholder text from the <p> element, as we do not need it anymore.

You can further generalize your code by iterating through every button you have. This way, you don't have to manually update your JavaScript code after you've changed some of the buttons. The code will adapt to the changes automatically. You can query every button with the `document.getElementsByClassName('mapchooser');` statement, which returns an array of DOM elements.

Creating web maps with Leaflet

Now we have everything in place except our maps. To create interactive maps, we can use external web mapping libraries written in JavaScript. One such library is Leaflet, which is perfectly capable of creating simple interactive maps. Leaflet has a simple and intuitive API, which is very easy to use. All we need is the library's code base and stylesheet loaded in our HTML document. Let's include Leaflet in our web page:

1. Download the stable version of Leaflet from `http://leafletjs.com/download.html`.

2. Extract the files in the downloaded archive in the web server's root folder. Optionally, create a new folder for the files to have a well-organized structure.

3. Edit the `map.html` file with a code or text editor.

4. Include Leaflet's code base (`leaflet.js`) with a `<script>` element using its relative path from the root folder in the HTML document's `<head>` section:

   ```
   <script src="leaflet/leaflet.js" type="text/javascript">
   </script>
   ```

5. Include Leaflet's stylesheet (`leaflet.css`) with a `<link>` element in the HTML document's `<head>` section:

   ```
   <link href="leaflet/leaflet.css" rel="stylesheet">
   ```

   You can also use the **CDN** (**Content Delivery Network**) used by Leaflet to include these resources. You can check out the correct references at `http://leafletjs.com/examples/quick-start/`. Note that using a CDN generates less traffic on your site, but takes some control out of your hands.

6. Leaflet's methods can be accessed with the `L` variable by default. Test the references by refreshing the web page, opening the developer tools, and writing `L` in its **Console**. If there isn't such an object, check the permissions of the extracted files. They must be readable by every user:

Elements   Console   Sources   Network   Timeline   Profiles   Application   Security   Audits   Layers   HTTPS Everywhere

🚫 ▽ top   ▼   ☐ Preserve log

\> L

‹ ▶ *Object {version: "1.0.3+ed36a04", Util: Object, Mixin: Object, Browser: Object, DomUtil: Object…}*

\>

Now that we have a reference to Leaflet's various methods, we can start creating web maps using its API. The API consists of methods exposed to the users of the library. Every good API has a thorough documentation on how to use the different methods, and Leaflet is not an exception. We can reach its API documentation at `http://leafletjs.com/reference-1.0.3.html`.

 If you need web mapping or Web GIS capabilities beyond Leaflet, and are proficient in JavaScript or willing to learn much more about it, OpenLayers is your library. You can get introduced to OpenLayers at `http://openlayers.org/`.

Creating a simple map

The first map we create should be the road map we created in GeoServer. We have to set up a blank map and load a WMS layer in it. Once it's working, we can bind the map to the appropriate button:

1. Create a new map with the `L.map` function and save it to a variable. It needs the ID of the container element as a string, while it can additionally accept other parameters bundled in an object. For now, let's provide a `center` and a `zoom` parameter. The center is an array of two WGS 84 coordinates (latitude and longitude), while the zoom is a positive integer. Supply center coordinates that fit our map. We can get them either from QGIS or from Google Maps by centering the map on our study area and noting down the coordinates from the URL:

```
var map = L.map('map', {
  center: [46.06, 18.25],
  zoom: 9
});
```

The map canvas is there, although it has some visual glitches. The problem is that Leaflet also uses <div> elements to create its containers, while we defined some styles on every <div> element. That is why creating styling rules that are too general can lead to broken design:

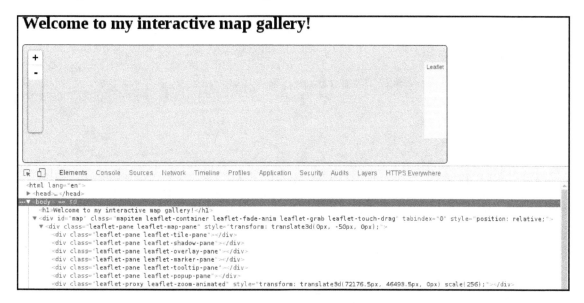

2. Edit the `map.html` file and define a class name on every <div> element (`class="mapitem"`).

3. Edit the `map.css` file and rename the `div` selector to the new class (`.mapitem`).

4. Now that our map canvas is fixed, we can load our first layer. We can create a new WMS layer with the `L.tileLayer.wms` function. It requires a mandatory URL parameter pointing at our WMS server and some optional WMS parameters (for example, `layers`, `crs`, `format`) bundled in an object. Let's just stick with the `layers` and `format` parameters. Create the WMS layer and save it to a variable. The `layers` parameter needs our layer group's name in GeoServer, along with the workspace's name it is in:

```
var wms = L.tileLayer.wms('http://localhost:8080/geoserver/ows', {
  layers: 'practical_gis:road_map',
  format: 'image/png'
});
```

5. Add the new layer to the map with the map's `addLayer` method. It requires a layer object as an only parameter. Use the variable containing the WMS layer:

```
map.addLayer(wms);
```

6. Inspect the result. Alter the map's `center` and `zoom` parameters, if required.

Leaflet supports chaining functions together. We can avoid saving our layer in a variable by calling its `addTo` method right after creating it. The method needs a map object as a parameter:

```
L.tileLayer.wms('http://localhost:8080/geoserver/ows', {
layers: 'practical_gis:road_map',
format: 'image/png'
}).addTo(map);
```

Our map is working, although we still need to bind it to our first button. We do not have to give up the general function we created, though. As object values can be anything, we can create a second object containing our buttons' IDs as keys and the functions creating our maps as values.

7. Create a new, empty object after the object, containing our descriptions:

```
var maps = {};
```

8. Define the `maps` object's first property. Name it after the ID of the first button. It should contain a function with the Leaflet code we created earlier. The function does not require any parameters:

```
maps.roadmap = function() {
  var map = L.map('map', {
    center: [46.06, 18.25],
    zoom: 9
  });

  L.tileLayer.wms('http://localhost:8080/geoserver/ows', {
    layers: 'practical_gis:road_map',
    format: 'image/png'
  }).addTo(map);
};
```

9. Add a third line to the `updateMap` function, executing the appropriate function from our `maps` object:

```
maps[evt.target.id]();
```

If we save our changes and refresh our page, we can see our road map by pressing the corresponding button:

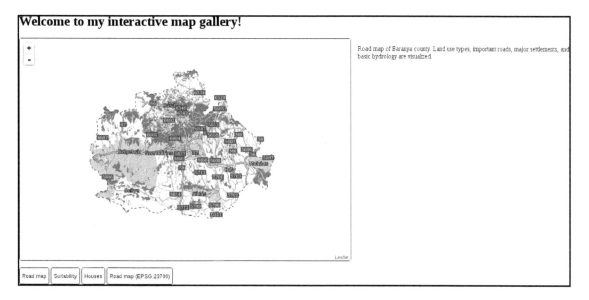

Compositing layers

For the next next task, let's create a map with our suitable zones. In this task, let's have an OpenStreetMap base map, the suitability layer from GeoServer as an overlay map, and the suitable zones as another overlay map:

1. Extend the `maps` object with a function. The key must be the ID of our second button. The function should have the same map initializing code that we used with our road map:

```
maps.suitability = function() {
  var map = L.map('map', {
  center: [46.08, 18.24],
  zoom: 10
  });
};
```

2. Add an OpenStreetMap base map to the composition. We can do that by calling the `L.tileLayer` function with the URL of the OSM tiles. As these tiles are not just separated by their layout and zoom level but also by cluster, we have to use wildcards according to the function's documentation:

```
L.tileLayer(
  'http://{s}.tile.openstreetmap.org/
  {z}/{x}/{y}.png').addTo(map);
```

The wildcards supplied in braces are automatically replaced to numbers by Leaflet. The `{z}` wildcard gets replaced by the zoom level's folder in the OSM server, the `{x}` wildcard by the row's folder, and the `{y}` wildcard by the tile's name, which represents the column's number.

3. Add our suitability map as a next layer according to the scheme we used with our layer group previously. In order to get transparent `NULL` values, provide an additional `transparent: 'true'` key-value pair in the list of optional parameters:

```
L.tileLayer.wms('http://localhost:8080/geoserver/ows', {
  layers: 'practical_gis:suitability',
  format: 'image/png',
  transparent: 'true'
}).addTo(map);
```

You can set any layer as transparent by providing an additional opacity value in the optional parameter list. By adding `opacity: 0.7`, you can set the opacity of the given layer to 70%. `0` means completely transparent, while `1` means completely opaque.

4. Add our suitable areas layer as the final overlay map using the same options we used in the previous layer, but alter the layer's name:

```
L.tileLayer.wms('http://localhost:8080/geoserver/ows', {
  layers: 'practical_gis:suitable',
  format: 'image/png',
  transparent: 'true'
}).addTo(map);
```

Now we should be able to see our second map in action:

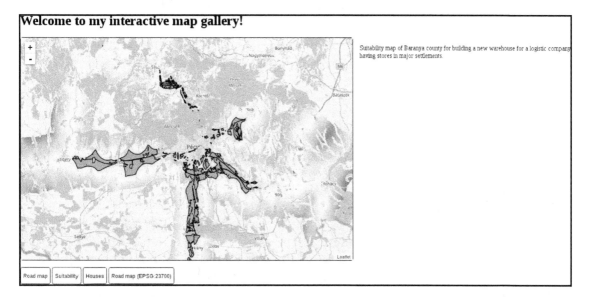

Now our second map loads, although we cannot change maps without reloading the page. The problem is that Leaflet recognizes if the provided container already has a loaded map and won't load another one until it is properly destroyed. We can destroy an existing Leaflet map; however, we need to understand another concept first--variable scope. In JavaScript, similarly to other programming languages, a variable cannot be accessed outside its valid scope. A valid scope in JavaScript is a function. We saved our map objects to variables in separate functions; therefore, they cannot see each other. Moreover, if we would like to access these map variables outside their functions, we would get an error. However, in these functions, we can access variables declared outside of them, just like we accessed our `maps` and `descriptions` objects. The solution is that we have to move our `map` declaration outside of the functions; by doing this, we can access the existing map instance (if any) before creating another one.

5. Declare the `map` variable in the first line of our code:

   ```
   var map;
   ```

6. Remove the `var` keywords prior to the `map` variables in every function using Leaflet. This way, we only assign map objects to the already declared variable and won't declare it again.

7. Write a statement in the beginning of the `updateMap` function, destroying the existing map. We can destroy a Leaflet map by calling its `remove` method. As we declared an empty `map` variable, it does not contain a map on the first click. Therefore, we should wrap our statement in an `if` statement. As empty variables default to the special value of `undefined`, we can provide a check against it in our conditional statement:

```
if (map != undefined) {
 map.remove();
}
```

Working with Leaflet plugins

One of the greatest strengths of Leaflet is its clean code base and great extensibility. The base Leaflet library is very small and lightweight; therefore, it has limited capabilities. However, a lot of developers have created a variety of interesting and useful plugins that can integrate into Leaflet, extending it to match our needs. The only weakness of this decentralized workflow is the lack of rigor and cohesion in plugins. Their quality (for example, documentation) can vary; therefore, we might need to fiddle quite a bit with some of them before getting meaningful results. We can check out a recommended list of plugins at `http://leafletjs.com/plugins.html`. Of course, there is no guarantee a listed plugin will be easy to use, although it is a good indicator of its quality.

Loading raw vector data

When it comes to web mapping and vector data, one of the most popular formats is GeoJSON. GeoJSON extends JSON, which is basically JavaScript's number one storage format. JSON is created to be concise while still able to represent almost every type in JavaScript. Therefore, we can save the states of web applications, for example, in order to resume them later. GeoJSON is also a concise format that can represent the basic geometry types and can be parsed by JavaScript easily. The only problem is that we have to load our GeoJSON files in our script. Although Leaflet knows how to handle GeoJSON objects, to load external content on the fly, we need to understand yet another concept--AJAX. On the other hand, there is a Leaflet plugin that can automatically load the content of a GeoJSON file with only URL.

Let's create our third map using this plugin:

1. Open QGIS and load the constrained houses layer. Export it with **Save As**. Choose the GeoJSON format and a CRS of WGS 84 (**EPSG:4326**). Name it `houses.geojson` and export it directly into the web server's root folder.

2. Using Leaflet's plugin list, look for the **Leaflet Ajax** plugin. The link will lead to the GitHub page where the code is maintained. Click on the releases tab, which will navigate you to the release page of the plugin directly at `https://github.co m/calvinmetcalf/leaflet-ajax/releases`.

3. Download the latest release by clicking on the **zip** link under it. The archive comes with the source files and the debug version, which we do not need. Extract only the `dist/leaflet.ajax.min.js` file next to the Leaflet library.

4. Edit the `map.html` file and include this new script in the <head> section after the Leaflet <script> element. As this is an extension, it needs Leaflet to be set up when it loads:

```
<script src="leaflet/leaflet.ajax.min.js"
 type="text/javascript"></script>
```

5. Create a new method for the third button in the `maps` object. The method should create a new map centered on the settlement we analyzed in Chapter 8, *Spatial Analysis in QGIS*:

```
map = L.map('map', {
  center: [46.08, 18.24],
  zoom: 12
});
```

6. Add an OpenStreetMap base layer to the map, just like we did previously:

```
L.tileLayer(
 'http://{s}.tile.openstreetmap.org/
{z}/{x}/{y}.png').addTo(map);
```

7. Add a new GeoJSON layer using AJAX with the `L.geoJson.ajax` function. It needs a relative path string as a parameter pointing to the GeoJSON file:

```
L.geoJson.ajax('houses.geojson').addTo(map);
```

If we click on the third button after refreshing the page, we can see our houses on the map:

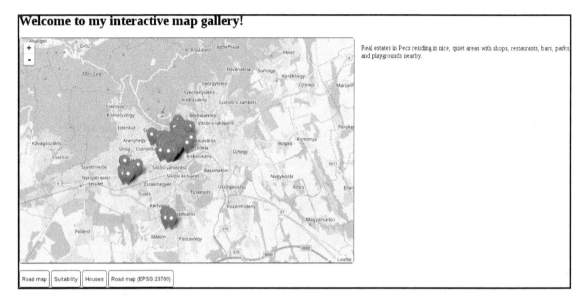

As we can see, Leaflet loads vector points as marker images by default. When we load some lines or polygons, it simply renders it as QGIS or GeoServer did. Let's load some polygons, but with WFS this time. Using WFS requests is another unsupported feature in Leaflet, which we can still use with another extension:

1. Look for the plugin called **Leaflet-WFST** in the plugin list of Leaflet's page. The link will open the plugin's GitHub repository, where the plugin can be downloaded in the **releases** page. Download the latest release by clicking on the **Source code (zip)** button.

2. Similarly to the previous plugin, we only need the minified version of the library and can discard the source files. Extract dist/Leaflet-WFST.min.js next to the Leaflet library in the web server's folder.

3. Include this plugin in the map.html file's <head> section after Leaflet:

```
<script src="leaflet/Leaflet-WFST.min.js"
  type="text/javascript"></script>
```

4. In `map.js`, remove the tile layer containing our suitable areas from the `suitability` method of the `maps` object.

5. Create a WFS layer instead. A simple WFS layer can be created with the `L.wfs` function. Unlike the previous functions, it only requires an object as a single parameter, and the object should contain the URL pointing at the OWS server with a `url` key. There are three more required parameters: `typeNS` with the queried workspace's name in GeoServer, `typeName` with the layer's name, and `geometryField` with the layer's geometry column. The default geometry column in GeoServer is `the_geom`:

```
L.wfs({
  url: 'http://localhost:8080/geoserver/ows',
  typeNS: 'practical_gis',
  typeName: 'suitable',
  geometryField: 'the_geom'
}).addTo(map);
```

If we display our suitability map, we can see the OSM base layer and the suitability layer, but no polygons. What happened? You might have already opened the developer tools for potential error messages and seen a message similar to the following:

```
XMLHttpRequest cannot load http://localhost:8080/geoserver/ows?.
No 'Access-Control-Allow-Origin' header is present on the
requested resource. Origin 'http://localhost' is therefore
not allowed access.
```

We stumbled upon two very characteristic concepts of web development--same-origin policy and **CORS (cross-origin resource sharing)**. The same-origin policy is a security measure that prevents browsers from mixing content from different servers. The path of the first document a web client loads determines the origin. From then on, the same-origin policy prevents the browser from loading anything that is not on exactly the same domain and exactly the same port. There are, of course, exceptions handled by CORS. These exceptions are typically scripts linked in `<script>` elements, stylesheets linked in `<link>` elements, images and other media, embedded content in `<iframe>` elements, and web fonts.

One of the things that the same-origin policy is always applied to is AJAX calls. WFS features, just like our GeoJSON file, are requested using AJAX, therefore the same-origin policy applies. The problem is that our GeoServer listens on and responds to port `8080`, while our origin is at port `80`. In the end, our web page cannot request resources other than images from GeoServer. The solution is simple; CORS can be enabled for any kind of resource on the server side using special headers in server responses. Enabling CORS in our GeoServer is slightly more complicated, though:

1. Edit the `webapps/geoserver/WEB-INF/web.xml` file in GeoServer's folder.
2. Look for the `cross-origin`, `<filter>`, and `<filter-mapping>` elements. There is a comment above them that states **Uncomment following filter to enable CORS**.
3. Uncomment the elements by removing the `<!--` and `-->` XML comment symbols around them.

 If you are using the web archive version from the Apache Tomcat Java servlet, there is only one thing left to do. You have to alter the value of the `<filter-class>` element in the `<filter>` element to `org.apache.catalina.filters.CorsFilter` if there is another value there.

4. GeoServer's Jetty version does not have the required code to apply CORS headers on responses bundled by default. We have to download and install it manually. Get the Jetty version used by GeoServer by looking in the `lib` folder of GeoServer (not `webapps/geoserver/WEB-INF/lib`).
5. The majority of the Java files starting with `jetty` contain Jetty's version number (for example, `jetty-http-9.2.13.v20150730.jar`).
6. The Java file containing the required code for CORS headers is called `jetty-servlets`. Download the `jar` file with the appropriate version from `http://central.maven.org/maven2/org/eclipse/jetty/jetty-servlets/`. For example, for version `9.2.13.v20150730`, you have to click on the corresponding version link and download the file named `jetty-servlets-9.2.13.v20150730.jar` by clicking on the file's link.
7. Copy the downloaded file to `webapps/geoserver/WEB-INF/lib` in GeoServer's folder and restart GeoServer.

8. After GeoServer runs, display the suitability map again to see our polygons:

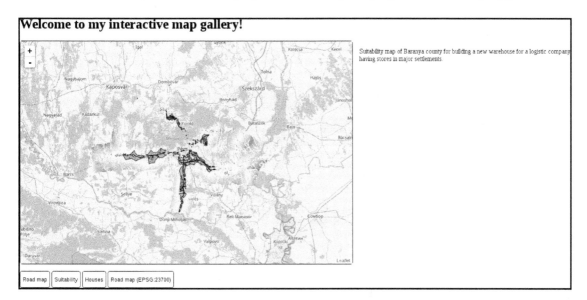

Styling vectors in Leaflet

As we can see, Leaflet represents vector points with marker images, while it draws lines and polygons just like QGIS does. The next thing we should learn is how we can customize the styles of the vector features we display in our web maps. Both the `L.wfs` and `L.geoJson.ajax` functions accept additional styling parameters we provide. Styling is done by passing a `style` object to the vector layer's function containing supported styling parameters. The complete list of parameters can be seen on the base vector class's API documentation page: `http://leafletjs.com/reference-1.0.3.html#path`. Using the properties in the provided list, let's change our polygons to better match the suitability maps we created in QGIS:

1. Edit the `map.js` file.
2. Extend the parameters of the `L.wfs` function with a `style` object. It should contain a `stroke` property with the stroke's color, a `weight` property with the stroke's width in pixels, and a `fillOpacity` property with a zero value:

```
L.wfs({
    url: 'http://localhost:8080/geoserver/ows',
    typeNS: 'practical_gis',
    typeName: 'suitable',
```

```
    geometryField: 'the_geom',
    style: {
        fillOpacity: 0,
        color: '#000000',
        weight: 1
    }
}).addTo(map);
```

By opening our sustainability map again, we should be able to see our new polygons:

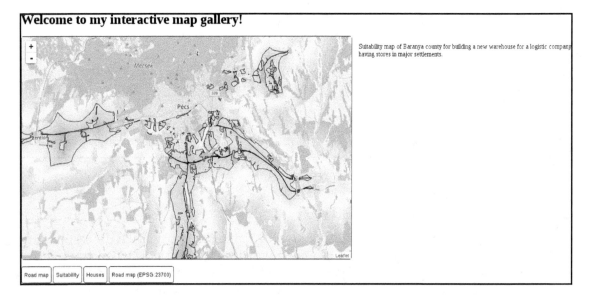

Customizing a single symbol for every vector feature in a layer is very easy in Leaflet. However, sometimes we would like to add thematics to our maps, such as attribute-based symbology. The `style` parameter of vector layers can aid us in this, as it also accepts a function. The function needs to be able to evaluate the style for every feature and return it in the end. Let's create a style for our houses map that visualizes our houses according to their prices:

1. Based on the price range we used in QGIS, define some intervals. For me, the ranges are `0-10,000,000`; `10,000,000-30,000,000`, and `30,000,000+`.
2. Supply a `style` function to the `L.geoJson.ajax` call in the optional object parameter. The property name is `style`, as usual. Style functions accept one parameter in Leaflet, the currently processed feature:

```
L.geoJson.ajax('houses.geojson', {
```

```
style: function(feature) {

  }
}).addTo(map);
```

3. In the `style` function, we should have only one variable: the fill color. Let's use three colors for our intervals--green, yellow, and red. Initialize a `fill` variable with the green color in the `style` function:

   ```
   var fill = '#00ff00';
   ```

4. We can override this color using conditional statements and checking the price attribute of the features against reference values. The easiest way is by checking whether the price is higher than the second interval's lower limit and then checking whether it's higher than the third interval's lower limit. The attributes of a feature can be accessed from its `properties` attribute, where they are stored in a simple object:

   ```
   if (feature.properties.price > 10000000) {
     fill = '#ffff00';
   }
   if (feature.properties.price > 30000000) {
     fill = '#ff0000';
   }
   ```

5. The next step is to create a `style` object and saving it in a variable. This is the `style` object that is returned by our function, and should contain every rule we would like to apply:

   ```
   var styleObject = {
     radius: 5,
     fillColor: fill,
     color: '#000000',
     weight: 1,
     fillOpacity: 1
   };
   ```

6. Finally, we should return the `style` object to Leaflet. We can define the return value of a function with the `return` keyword:

   ```
   return styleObject;
   ```

Creating a Web Map

If we reload our houses layer, it still contains the usual blue markers. The problem is that vector points are visualized as marker images, and marker images cannot be styled as vector features. We have to override this default behavior of Leaflet and visualize our vector points as circles. There are two functions we can use for this. The `pointToLayer` property can be supplied on point layers to specify how Leaflet should draw the vector point. It has to be a function with two parameters: `feature` and `coordinates`. The `coordinates` parameter passes the WGS 84 coordinates of the processed point directly; therefore, we can easily create another symbol from them. The function must return the vector object that can be directly used by Leaflet to draw our point. We can create a stylable circle with the `L.circleMarker` function, which only requires two WGS 84 coordinates.

7. Provide a new property to `L.geoJson.ajax` besides `style`. Its name must be `pointToLayer`, while its value must be a function with two parameters. The parameter names are not fixed, but their order is:

```
pointToLayer: function(feature, coords) {

}
```

8. In the function, simply return the result of an `L.circleMarker` function called directly with the `coordinates` parameter:

```
return L.circleMarker(coords);
```

The initialization call of our GeoJSON layer now should look like the following:

```
L.geoJson.ajax('houses.geojson', {
  style: function(feature) {
    var fill = '#00ff00';
    if (feature.properties.price > 10000000) {
      fill = '#ffff00';
    }
    if (feature.properties.price > 30000000) {
      fill = '#ff0000';
    }
    var styleObject = {
      radius: 5,
      fillColor: fill,
      color: '#000000',
      weight: 1,
      fillOpacity: 1
    };
    return styleObject;
  },
```

```
        pointToLayer: function(feature, coords) {
          return L.circleMarker(coords);
        }
    }).addTo(map);
```

We can inspect our new thematic house map by refreshing the page and loading the map:

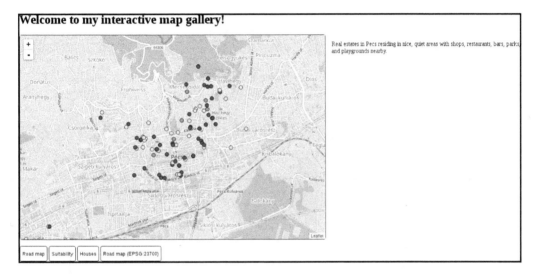

Annotating attributes with popups

A truly great feature of Leaflet is its convenient popup management. We don't have to create any kind of popup mechanism (HTML elements, close button, or custom code on different events); Leaflet can create full-fledged popups automatically just by providing an HTML string. An HTML string is a simple string that can contain HTML tags. The HTML tags are interpreted rather than printed directly, and the resulting DOM structure is rendered instead. Let's create some popup content for our houses layer:

1. Add an additional property to the optional list of properties of the `L.geoJson.ajax` layer. The property name must be `onEachFeature`. It requires two properties: one for every processed feature and one for the processed feature's layer object, which has the `bindPopup` method. The function can be void; therefore, it does not need a return value:

   ```
   onEachFeature: function(feature, layer) {

   }
   ```

2. In the function, first, create an HTML string and save it into a variable. Different strings can be concatenated with the + operator. When a number is **added** to a string, the number is automatically converted into its string representation. Use three attributes of the features--the `name` containing the street name, the `size` containing the size of the house, and the `price` containing the price of the house. Additionally, use the `<br>` tag to insert line breaks:

```
var htmlString = 'House on ' + feature.properties.name +
  '.<br>Size: ' + feature.properties.size + ' m²<br>Price:
  ' + feature.properties.price + ' HUF';
```

3. Bind a popup to the feature using the layer object's `bindPopup` method. It requires only an HTML string as a parameter:

```
layer.bindPopup(htmlString);
```

Our house map now annotates some of the attributes of our houses:

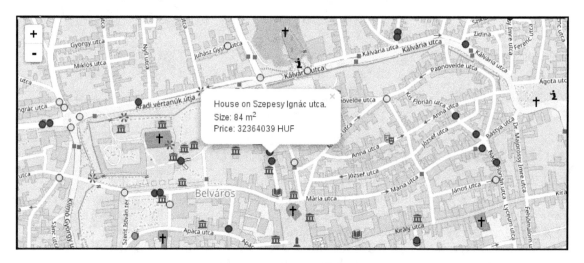

 One of the potential problems with using third-party plugins to extend Leaflet is the varying implementation quality. For example, `L.geoJson.ajax` extends the base vector layer in a way that its original methods, such as `onEachFeature`, can be used. On the other hand, `L.wfs` does not make this possible. Popups can still be registered to WFS features in Leaflet, although it needs a better understanding of the library.

Using other projections

Leaflet can use some projections out of the box, of which two have cartographic significance--EPSG:3857 and EPSG:4326. The default projection is the Web Mercator, while the WGS 84 projection has to be specified in the map's `crs` property. The `crs` property's value must be a valid Leaflet `projection` object. This object contains methods to transform coordinates around and has predefined zoom levels fitting the given projection, among other things. Let's try out the WGS 84 projection on our road map first:

- Extend the `maps` object with a function corresponding to the ID of our fourth button. The function should be almost the same as the other road map function. The only difference is that the map should contain an additional `crs` parameter with the `L.CRS.EPSG4326` projection object:

```
map = L.map('map', {
  center: [46.06, 18.25],
  zoom: 9,
  crs: L.CRS.EPSG4326
});
L.tileLayer.wms('http://localhost:8080/geoserver/ows', {
  layers: 'practical_gis:road_map',
  format: 'image/png'
}).addTo(map);
```

To use other projections, we can use yet another plugin. This plugin can utilize PROJ.4's JavaScript port--`Proj4.js`. It can create regular projection objects that Leaflet can use from projection definitions. Let's create a projection object representing our local projection:

1. Look for the plugin named `Proj4Leaflet` in Leaflet's plugin list. The link points to its GitHub page, where we can download the latest release from the project's `Releases` page.
2. We need two additional libraries, as `Proj4Leaflet` uses and depends on `Proj4.js`. Extract `lib/proj4.js` and `src/proj4leaflet.js` from the downloaded archive next to the Leaflet library.
3. Edit `map.html` and include the two libraries in the <head> section, as usual:

```
<script src="leaflet/proj4.js" type="text/javascript">
</script>
<script src="leaflet/proj4leaflet.js" type="text/javascript">
</script>
```

4. Look up the PROJ.4 definition of our local projection. We can use `http://epsg.i` `o` for this. Search for the EPSG code of our local projection, and on the projection's page, select **PROJ.4** from the **Export** menu. Copy the definition string to the clipboard by clicking on **Copy TEXT**.

5. Go back to `map.js`, and in the function of the fourth map, create a new projection with the `L.Proj.CRS` constructor function available now. Constructor functions are special functions used to instantiate classes. For now, it is enough to know that they have to be called with the `new` keyword. The constructor function needs three parameters--a projection name containing the EPSG code of the local projection (for example, `EPSG:23700`), the PROJ.4 definition string of the projection, and an object containing optional parameters. The only optional parameter we need to provide is an array of resolutions that Leaflet will use for the various zoom levels in a descending order. There are no rules for the values of this array, but as a rule of thumb, using the powers of 2 (for example, 2, 4, 8, 16) is a good choice. We can start from `4096` and increase it later if it is not sufficient:

```
var localProj = new L.Proj.CRS('EPSG:23700',
  '+proj=somerc +lat_0=47.14439372222222
  +lon_0=19.04857177777778 +k_0=0.99993 +x_0=650000
  +y_0=200000 +ellps=GRS67 +towgs84=52.17,
  -71.82,-14.9,0,0,0,0 +units=m +no_defs',
  {
    resolutions: [4096, 2048, 1024, 512, 256, 128, 64, 32,
      16, 8, 4, 2, 1]
  });
```

6. Supply the variable containing the new `projection` object in the `map` function's `crs` parameter. We don't have to modify the center coordinates, as Leaflet needs them in WGS 84. It transforms them automatically if it needs to do so. We might have to adjust the starting zoom level, though, as we are using a custom set of resolutions:

```
map = L.map('map', {
  center: [46.08, 18.24],
  zoom: 4,
  crs: localProj
});
```

Our fourth map is now working, and showing our road map in our local projection:

Welcome to my interactive map gallery!

Road map of Baranya county in EOV projection. Land use types, important roads, major settlements, and basic hydrology are visualized.

Road map | Suitability | Houses | Road map (EPSG:23700)

Automatic coordinate transformation on the client side does not restrict to transforming the center coordinates of our maps. It is a very handy property of web mapping libraries to save valuable server resources.

Summary

Congratulations! You just created several different web maps using different techniques. In this chapter, we learned how the client side of the web works and how we can utilize a web mapping library for our mapping needs. We can now use Leaflet to easily share our maps on the Web. We can load image layers, vector layers, style vector data, popups, and even custom projections. We also discussed some of the basic concepts of the frontend, such as the interaction of HTML, CSS, and JavaScript elements, or CORS.

This is the end of our journey. We learned a lot of things about GIS from the theory behind data models by creating a spatial database to publish the results on the Web. We did our best to understand how spatial data works and how it can be analyzed through various means. We got some meaningful results from our example data, which is completely open source (in most cases), just like the software we used. Take a good rest, but don't forget--this is only the beginning of a long journey if you wish to master the art of GIS. I hope you will not only apply this knowledge to practical problems, but will also find joy in doing so.

Appendix

Appendix 1.1: Isosurfaces (800 mm and 1200 mm precipitation) visualized on the Digital Elevation Model of Slovakia provided by the sample slovakia3d dataset for GRASS GIS:

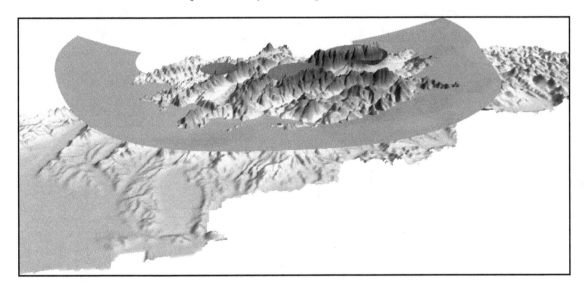

Appendix 1.2: The same map we created in `Chapter 2`, *Accessing GIS Data With QGIS*, only with a CRS using an Albers Conic projection (EPSG:102008). The map is not North-aligned; therefore, a north arrow was used with the **North alignment** parameter set to **True north**:

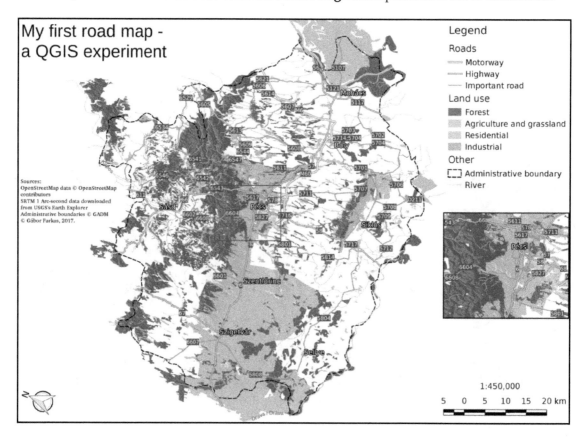

Appendix 1.3: Some of the basic geoalgorithms visualized. a: the two input layers (A and B per GRASS's **v.overlay**), b: clip (and), c: union (or), d: difference (nor), e: symmetrical difference (xor):

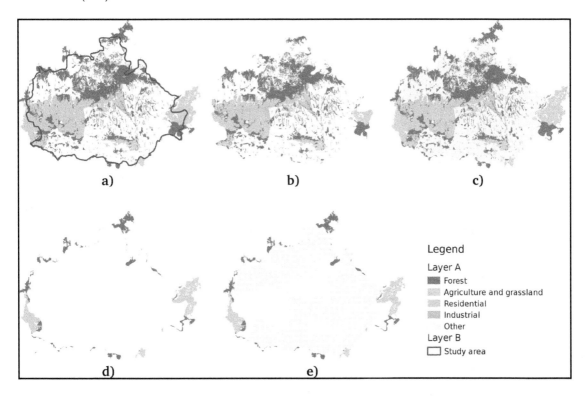

Appendix 1.4: Some of the Coordinate Reference System's PostGIS support. The selected one is the **EPSG:4326** CRS used by the book in the early chapters:

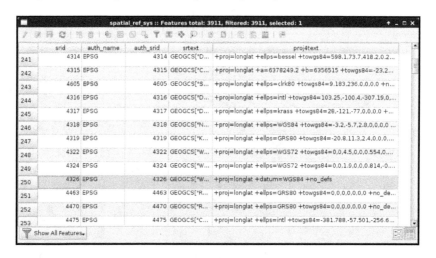

Appendix 1.5: Settlement data of my study area downloaded from the OpenStreetMap database via the QuickOSM plugin using the Overpass API:

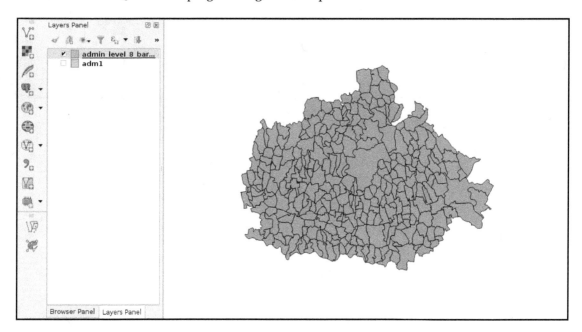

Appendix 1.6: Linear interpolation of point v_i on the line segment between points v_0 and v_1 based on a factor f. If v_1 - v_0 divided by $1/f$ leaves a remainder, the interpolated coordinate needs more precision (left). If not, the interpolated coordinate will have the same precision (right). In GIS, however, it is a common practice to store coordinates with a fixed precision in a single vector layer:

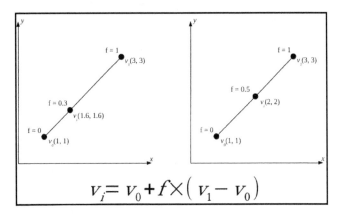

$$v_i = v_0 + f \times (v_1 - v_0)$$

Appendix 1.7: Difference between precise proximity analysis with buffering (green points), and selecting features with precision values (yellow points) in QGIS. Every green point is selected, but there are some points excluded from the precise results:

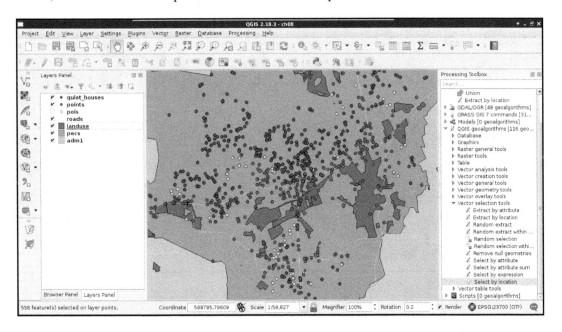

Appendix 1.8: Do you need an **Extract by expression** tool in QGIS? Build one yourself! You can easily create such a model by requiring a **Vector layer** input of any type, a **String** input for the expression, and linking the **Select by expression** and the **Save selected features** tools together:

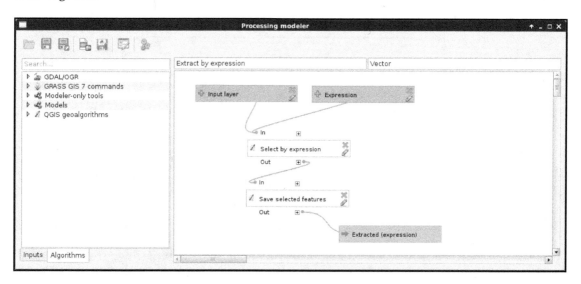

Appendix 1.9: Interpolated surface from contour lines visualized in 3D with the contour lines displayed on the interpolated surface:

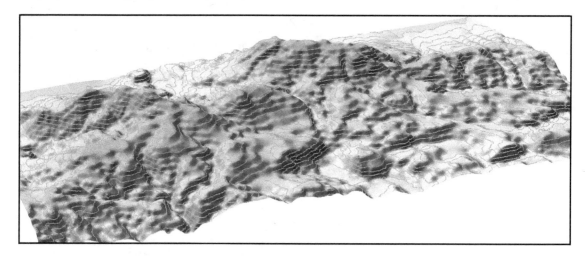

Appendix 1.10: The same vector layer transformed to raster with a resolution of 5 meters (left) and 2 meters (right). By using 5 meters, two of the possible houses overlap with buildings; thus, get cut off roads in the walking time analysis:

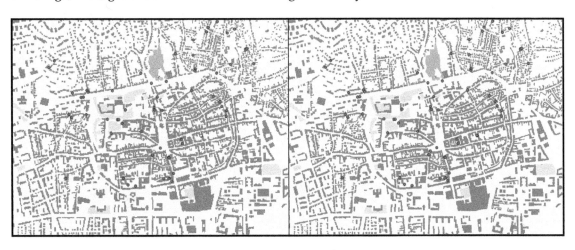

Appendix 1.11: Buffer zones with a different number of segments are used for more precise approximations. The yellow point lies outside of the buffer zones created with **5** and **10** segments; however, it is only a matter of precision as it is located inside the zones as proven by the buffers created with **15** and **50** segments:

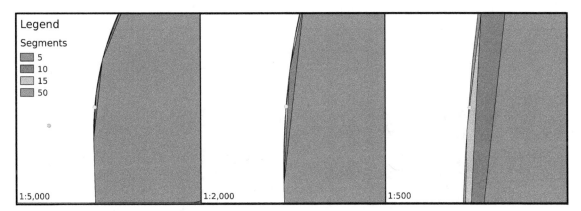

Appendix 1.12: Three popular distance types in GIS. In the Manhattan distance (left), we can go only in one dimension at a time; in the euclidean distance (center), we can move in two dimensions at a time; while in the case of the great-circle distance (right), we can move in three dimensions at a time, but only on the surface of a sphere. Note that the Manhattan distance between the same pair of points does not change, no matter how many breaks (turns) we use:

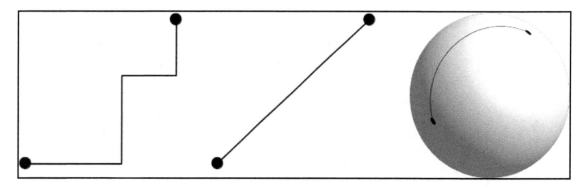

Appendix 1.13: Using clipped layers with NoData values in GDAL's **Proximity** tool introducing edge effects (left). NoData values are reclassified to zeroes, which wouldn't be a problem; however, distances are calculated from the edges, like they were features. It does not matter if we clip the distance matrix again (right), the edge effects are already introduced to the analysis:

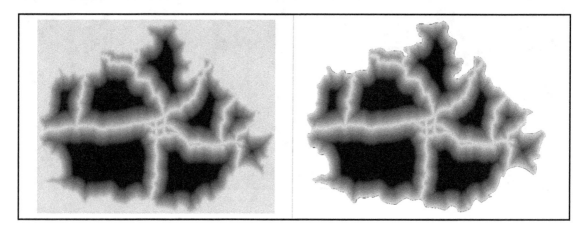

Appendix 1.14: Other popular fuzzy membership functions. Some of them are similar to the ones described in `Chapter 10`, *A Typical GIS Problem*, although with slightly different shapes and different formulas. The formulas were created with the assumption that the whole data range is fuzzified. If not, cell values below or above the threshold should be handled. In the formulas, variables denoted with `m` are break points or turning points, while variables denoted with `s` are defining the shape of the functions.

The natural number `e` is not included in QGIS's raster calculator, although, similarly to π, it can be hard coded as 2.7182:

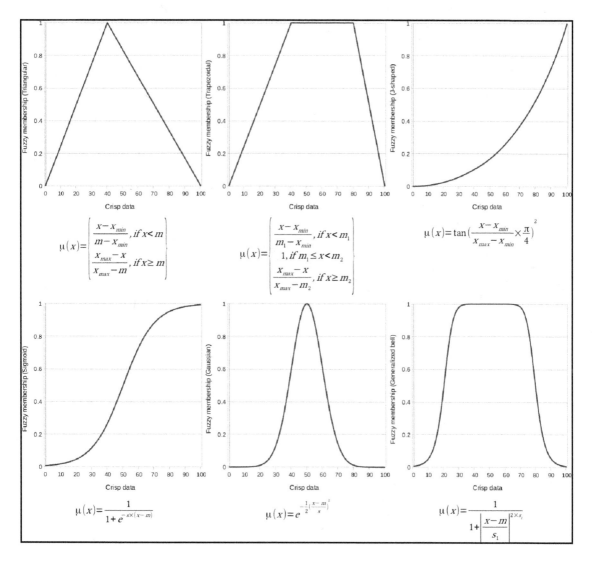

Appendix 1.15: Using an OpenStreetMap base layer under the suitability map. The suitability map has a transparency of 30%:

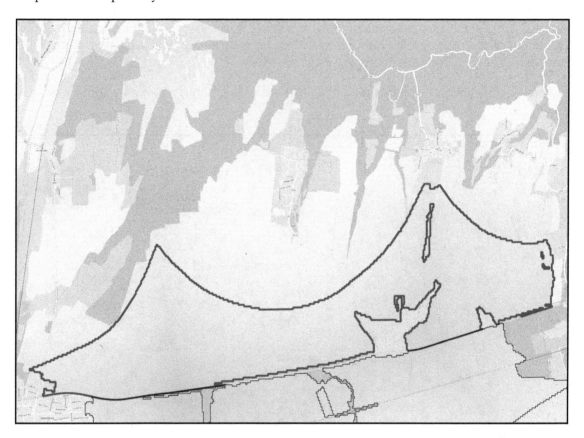

Appendix 1.16: A server machine hosting two web servers (one for running Java web applications), a PostgreSQL database, and an SSH server. Clients can query those servers with the appropriate client-side applications:

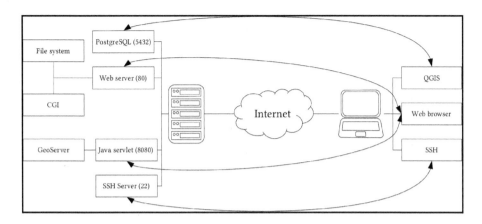

Appendix 1.17: Hillshaded land uses polygons in GeoServer. You can achieve the same by using **Raster** ∣ **Analysis** ∣ **DEM** in QGIS with the default **Hillshade** mode to create a static relief raster. Then, you can clip the relief to the area of the polygons using **Raster** ∣ **Extraction** ∣ **Clipper** with the option of cropping the result to the cutline. Finally, you have to load the resulting raster into GeoServer, and blend it into the land use layer using an overlay composition:

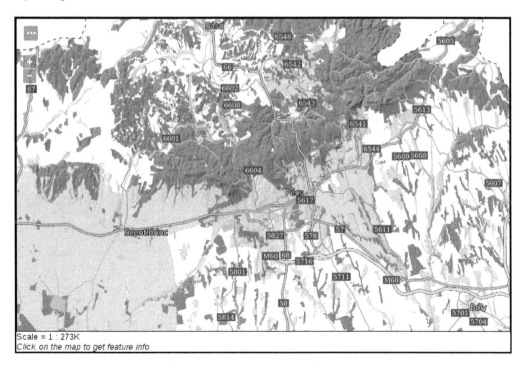

Appendix 1.18: Common line cap and line join styles used in vector graphic software:

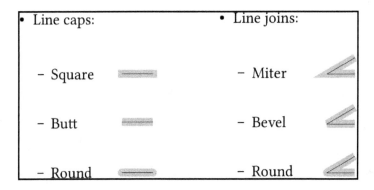

Index

www.ingramcontent.com/pod-product-compliance
Lightning Source LLC
LaVergne TN
LVHW081328050326
832903LV00024B/1077